EDITOR'S NOTE

While we were driving from Rye to Hastings in July of last year my wife noticed a car boot sale in a field by the road and we decided to stop for a look. On the stall of two brothers who make a living from house clearances in the area I spotted Sherlock Holmes's famous *Practical Handbook of Bee Culture*. Hardly able to believe my good luck, I bought it for £2.

The book records Holmes's bee-keeping activities while in retirement in East Sussex. It has always been thought that no copies of this legendary volume had survived, one of which Holmes gave to Watson on the eve of World War I, as recorded in the story *His Last Bow*. To my amazement, I found that the *Handbook* is not just about bee-keeping. Holmes includes in it material about his day-to-day life in Sussex – including his marriage to Mrs. Hudson – as well as stories of investigations he undertook during the period 1904-1912, and accounts of two terrifying attempts that were made on his life. The revelations in some stories are so indiscreet that I believe Holmes had the book privately printed for circulation only among his intimate circle.

I'm pleased to be able to make the text of the *Handbook* generally available; I'm sure it will delight all followers of the great detective. And it contains a quite extraordinary treasure — several photographs taken by Holmes himself, four in which he actually appears, two showing Dr. Watson and one showing the recently-married Mrs. Holmes.

Paul Ashton, March 2017

Hardcover ISBN 978-1-78705-123-2
Paperback ISBN 978-1-78705-124-9
ePub ISBN 978-1-78705-125-6
PDF ISBN 978-1-78705-126-3

Published in the UK by MX Publishing
335 Princess Park Manor, Royal Drive, London, N11 3GX
www.mxpublishing.co.uk

PRACTICAL HANDBOOK
OF
BEE CULTURE

WITH SOME OBSERVATIONS
UPON THE SEGREGATION OF THE QUEEN

SHERLOCK HOLMES

PRINTED BY
W. W. WASS & SON
60 CAIRO ROAD, LEWES
1913

Dedicated to

Baroness Angela Burdett-Coutts (1814-1906)
the great philanthropist of the age,
first female Member of the Royal Society
and
President of the British Bee-Keepers' Association 1876-1904

PREFACE

If you start a vineyard in your garden, you cannot hope to produce wine equal to that of the great French vineyards. With honey, there is no such disadvantage. In each hive, there will be thirty thousand experts working with you, and their family has had a hundred million years of experience in the business. You are in good hands.

I had this thought in mind when, in 1903, I retired from active life as a consulting detective in London and came to live in this former farmhouse in Sussex. The age of forty-nine might be thought somewhat early for retirement, but the state of my health dictated it.

I decided to keep bees from the start, and since coming here I have kept a journal of my bee-keeping ups and downs; this book is a selection from the entries in my journal over eight years. I publish it in hope that it may encourage others in my circle to take up bee-keeping and to discover, as I have done, the pleasures of this most rewarding of occupations.

Readers must not think, however, that this book is a cyclopaedia of everything pertaining to bees, like *The ABC and XYZ of Bee-Culture*, by A.I. and E.R. Root of Medina, Ohio, or *Practical Bee-Guide: A Manual of Modern Bee-Keeping*, by the Rev. Joseph Digges of Mohill, County Leitrim. I have nothing but praise for such works, but there is no denying that those magnificent and indispensable tomes – which I daily consult – are somewhat dry as reading matter. I have taken a different course. I hope my readers will not feel deceived, as they enter on these pages, to discover that into my bee-journal I have interpolated many entries about my life lived away from the hives.

There are also reminiscences of some of my cases which Dr. John Watson mentions in his now widely-known stories, but does not expand on there, and four accounts, not previously published, of my involvement in cases that have attracted some public attention, namely: the Whitechapel murders; the affair of Dr. Crippen; the Latvian anarchist outrage which ended in Sidney Street, London E.; and the theft of *La Gioconda* from the Louvre in Paris. In addition, I have included the stories of four investigations undertaken in the last seven years, two of which very nearly proved fatal to me.

In January 1904, I became the proud owner of one of the first Box Brownie cameras. I have illustrated my text here and there with photographs taken with it, and with some other photographs taken by others.

Finally, I should like to record here my gratitude to my wife for the nine years of love and companionship we have shared since we moved to this most beautiful spot.

Old Home Farm
East Dean and Friston, Sussex
November 1912

1903

Sunday, 1st November In the first week of October I bought a hive of pure Italian bees for 15/- from Mr. William Tompsett of Friston Forest. Provided the colony is a strong one, he told me, one need not worry overmuch about what kind of bees they are, but Italians are good for a beginner such as myself. The weather has been so good this Autumn that the bees are still active, and have not yet gone into their winter hibernation. Later that month I acquired three more colonies, all Italians and very healthy, two from Captain Wickham-Jones, of Jevington, and one from Miss. J. M. Battalons, of West Dean.

I placed my four hives under the fruit trees at the end of the garden, facing south-east to catch the early rays of the sun. This orientation also stops the prevailing south-westerly wind blowing in through the entrances.

I was immediately struck by the care with which the bees marked the location of their new home. On the afternoon of Wednesday, October 28th, I settled myself in the grass under the copper beech and watched them for two good hours, sitting as motionless as the Great Buddha at Kamakura. At first, they flew only a few inches from the entrance and passed back and forth many times, always facing the hive. Each time they gradually lengthened their line of flight, back and forth, up and down, until they received an indelible impression of the appearance of their home. After fully examining the front of the hive they flew a little further until they could get a similar view of the immediate surroundings at a distance of a few feet. Their flight then took the form of irregular circles, gradually enlarged to take in the apiary area and in time the whole of the garden, the orchard and the surroundings. After this thorough preparation, they can now fly directly to the hive entrance from any point of the compass.

Mr. Jas. Hiams, my next neighbour and one of the most experienced bee-keepers in the county, told me that a bee rarely flies more than three miles from the hive. He said this in a tone somewhat apologetic, I thought, about this limitation. A creature half an inch long, heavily laden with pollen on the return journey, often buffeted in these coastal parts by wind, rain and hail, and needing to navigate her way home over fields and downs with absolute precision! I am lost in admiration of her. Three miles seems to me astonishing, in the circumstances.

Sunday, 8th November We are now pretty well settled into this house, and though profoundly rural, we do not lack for metropolitan comforts. I write by a gas light that has the latest in incandescent mantles. Air is mixed with the gas and passed through muslin treated with zirconia and other minerals, to provide a brilliant, steady light. Yes, I know there are now two and a half million electric lights in London, but we are happy with the gas. I have most of my Baker Street paraphernalia about me, except my stimulants of old, of course. And, now I think of them, I cannot find the Persian slipper for my tobacco, and my mouse-coloured dressing gown, both of which which seem to have got lost in the move from London. They may yet turn up in one of the still unopened boxes. When I look up from my writing I see, on the cabinet opposite, the bust that was shattered with such skill by Colonel Sebastian Moran in early April, 1894.

I also have my casebooks, awaiting their shelves, and my bulging scrapbooks of crime report, and my unrivalled collection of the

Penny Dreadfuls of the century just gone. I have been adding an entry in a scrapbook today. On Thursday, 5th April, 1900, sixteen-year-old Jean-Baptiste Sipido stood with a revolver in his hand and his pockets crammed with anarchist literature, watching the Prince of Wales board a train at the Brussels Noord Station. As the train pulled out, Sipido jumped onto the footboard of the Prince's carriage and fired twice into the saloon. He missed, and was dragged off the train by station staff. I have recently learned that he gave as the reason for his action his wish to avenge the thousands killed in the South African wars, and have added that note to his entry.

My wife-to-be has done sterling work these last two months with wall-paper and *nuanciers* from Morris and Co., druggets, chequered linoleum, French hempen sheets, an ottoman, mahogany furniture – in a taste somewhat less modern than mine, but so be it – and, wonder of wonders, a Grand Rapids carpet sweeper. My own responsibilities have been: overseeing the restoration of the downstairs fireplaces, all now draped and over-mantled satisfactorily; the installation of a cast-iron bath with clawed feet, which took six strong men from Eastbourne to get it off the wagon and up the stairs, along with a marble-topped wash-stand and many panels of frosted glass; and arranging delivery of the kitchen dresser and the coal-fired kitchen range. I feel an affinity with the range; like myself, it is somewhat temperamental.

Where there is no wallpaper we have mainly yellow or buff walls with a dado and woodwork of chocolate, or olive brown. We have ceilings of straw colour, or pink, having banished the cold white that was there when we moved in. In the front of the house we have Venetian blinds of buff union cloth, and at the back the same thing in red.

The house abounds already in capacious closets and cupboards, but we have bought three mahogany wardrobes, two of them japanned. My papers, including my alphabetical list of criminals, which I am revising, are confined to this study at the back of the

5

house where I am now. It is very much the same size as my sitting room in Baker Street, and overlooks our garden and little orchard.

We have a cook and housekeeper, Mrs. Trench, and a housemaid, Pearl Thomas, both from Brighton, who live in, and a gardener and tree man, Bill Frusher, who lives in the village.

Friday, 20th November Wonderful news. I had a letter today from Paris, from my friend and correspondent M. Henri Becquerel. He and his colleagues, Pierre and Marie Sklodowska Curie, have been awarded the Nobel Prize for Physics. I have been in correspond-ence with M. Bequerel since 1896 on the use of photographic plates in detecting emissions from phosphorescent uranium salts.

The award is in recognition of their researches into the phenomena of radiation. I heard M. Curie speak on this subject in London in June of this year, at the Royal Institution. By the rules of the Insti-tution, Mme. Curie was not allowed to address us, because she is a woman. I stood up at the start of the proceedings and protested against this absurdity, but, though I was listened to in silence, there was only a scattering of applause when I sat down, and no comment from the Chair. Well, now she is the first woman in the world to receive a Nobel Prize. I hope that has caused some coughing and spluttering, and perhaps even an apoplexy or two – one can but hope – in Albemarle Street. Pierre Curie showed how radium rap-idly affects photographic plates wrapped in paper, how the sub-stance gives off heat, and in semi-darkness he demonstrated its sin-ister glow. He described how he wrapped a sample of radium salts in thin rubber and kept it strapped to his arm for ten hours. It caused a wound very like a burn, and we saw the grey scar that re-mained.

Saturday, 28th November Tomorrow I am going down to Birling Gap to take my first swim in the Channel. A challenge to the strong-est frame, but I imagine I will survive. I hope to make it a weekly commitment. My bees are more prudent. They have clustered themselves together in their hives, where they will lie dormant till the Spring. There is very little for bee-keepers to do at this time of year except to meet and swap yarns. I shall be attending a lantern-slide lecture on "The Enemies of the Bee" next week, in East-bourne.

Feather beds are not now, I am told, in such general use as for-merly, but my wife-to-be is in favour of them. We have six of them in the house now, at about seven pounds each. I have become something of an expert on feather beds, and am considering a mon-ograph on the subject. There are various qualities of goose feathers, distinguished by different names, though to the uninitiated they ap-pear very nearly alike in everything but colour. The best feathers are fluffy, with down on the stems, and are curved, or *curled,* as it is termed. The fluffier, the better. The best white feathers have fluff-iness in perfection; they are also cleaned and bleached. A good bed may easily be recognised if, on pressing it, the feathers rise quickly, forcing the ticking up with them. There is no doubt that spring mat-tresses will, when they can be cheaply made, supersede feather beds, as being less liable to take infection, more easy to arrange, and occasioning less dust.

Monday, 14th December Preparations for Christmas have begun in earnest. Fiona's brother, George Moralee, his wife and their three children will be coming from Carlisle to join us. My dear Wat-son has his own family ties, so cannot be with us, but he will be here for the special event we have planned for next January. We shall also be entertaining Jas. Hiams and his wife, our neighbours to the east, along with Mrs. Hiams's sister, Martha, and Mr. and Mrs. Scrope Viner, of Friston, and their two children.

I will permit myself an occasional, purely personal, reflection in this journal. This morning I saw my face reflected in a shiny red glass ball as I was tying it to the branch of our Christmas tree in the down-stairs sitting room, and had an unexpected thought. I was amazed to realise how different I am now from the man I was, only a year ago. I could never have imagined, in Baker Street, decorating a tree, or wrapping Christmas presents. I said to myself: Holmes, perhaps one *can* teach an old dog new tricks. This thought cheered me as we got ourselves thoroughly entangled in paper chains, greenery laced with red ribbon, and the high mysteries of the traditional plum pudding, which I was permitted to observe from afar.

Saturday, 26th December Yesterday we dined on a leg of pork, with the aforementioned pudding to follow, which, to applause, I carried in, wreathed in blue brandy flames. The pudding wreathed in flames, that is, and not myself. For supper we had cold salt beef and brawn, with pickles, followed by marzipan, muscatels and coconut Turkish Delight – called by Lydia Scrope Viner "lumps of delight". Today we walked a good six miles down to the cliffs and back, the Moralee children delightful and enjoying the clear air, but very cold. This evening were charades, an entertainment new to me, in which one acts out in dumbshow the titles of books, plays, popular songs, and what not. I gave them "Treasure Island", and Henty's "The Dragon and the Raven". Well, my first Christmas *en famille*, as one might say, has gone pretty well. A new life, in the new year, awaits.

Wednesday, 30th December I have had some time these last two days to commit to paper my understanding of what are the principal races of bees. Needless to say, as an absolute novice of only a month's experience I owe what follows principally to my bee-mentor, Jas. Hiams, now in his sixtieth year of bee-keeping. I hope it may be useful to other amateur apiarists like myself.

Italians Italians, sometimes called Ligurians, together with their crosses, or hybrids, are generally recognised as the most valuable, amiable and resistant to disease of all bee strains. They were introduced into this country in 1859 from a district in the Alps including portions of Switzerland and northern Italy. They are more prolific than the Blacks, more active, ready for swarming earlier, and can collect nectar from flowers where the Blacks cannot. There are considerable differences in the appearance of the various strains. The 'three-banded' are usually regarded as the most desirable, with three distinct yellow rings about the body below the wings. The 'goldens' are also highly regarded.

Cyprians Cyprians, originally from the island of Cyprus, resemble Italians. They are prolific, and are diligent workers, but are much more difficult to control than are the Italians. In fact, they are so cross and vindictive that I understand most bee-keepers have discarded them, and they are seldom offered for sale.

Syrians Syrians are similar in this respect; they are wicked, most difficult to handle and are often quite unmanageable.

Common Black, or German Common Black, or German, bees were the first stock imported into America but are now found there only mixed with Italians. They are not so gentle as Italians, and neither do they resist disease so successfully, but they begin working and breeding early, and are excellent comb builders.

Carniolans Carniolans are natives of Austria, resemble Common Blacks except for white bands on the lower abdomen, are peaceable, like the Italians, and have prolific queens, but their propensity for excessive swarming makes them unpopular.

Caucasians Caucasians are natives of the Caucasus, in Russia. Within the past few years they have been recommended in the United States by the Department of Agriculture, and testimony to their exceeding gentleness and prolificness has been given by many prominent bee-keepers there. Neither smoke nor protection is necessary, apparently, when these bees are being handled; they show little resentment when roughly treated, their queens are prodigious layers, and their workers exceptionally industrious.

(I say nothing here of the fearsome Giant Bee of India, native also of Ceylon, China and Java; of the Common, and the Dwarf, East Indian Bee; the stingless bee of South America; nor of the solitary

9

Sand Bee and Leaf-Cutter Bee. Interesting though they are, they are unlikely to cross the amateur bee-keeper's path in this country.)

1904

Friday, 1st January Last night Fiona and I took part in a capital concert given in the Peel Ward of Eastbourne General Hospital, for the entertainment of the patients. There was a large attendance, with over 100 patients and visitors present. The ward was decorated with foliage plants, bunting and artistic lamps.

This was the programme:

The Dolly Suite, by Gabriel Fauré (Pianoforte duet): the Misses Luxford and O'Flahertie
Humorous recitation: Dr. Jawalhir Lal Sinha
Chants de Tendresse et de Gaieté (Vocal performance): Mrs. Axel Hudson, Sister Angela Milne, Dr. Norman Gibbs (House Surgeon), Dr. Bertram Ground, and the Rev. W. L. Chick
Comic duologue: Sister Lily Simpson and Staff Nurse Martha Sparkes
Caprice No.24 for solo violin, by Niccolò Paganini: Mr. Sherlock Holmes
General choral singing, led by Dr. Martin Wrigley (*A Bird in a Gilded Cage; Bill Bailey, Won't You Please Come Home; Meet Me in St. Louis; Goodbye, Dolly Gray; Sweet Adeline; Auld Lang Syne*)

This morning Jas. Hiams came over and put a cake of what he calls bee fondant in each of my hives. He says it is a tradition in these parts to give the bees a New Year's gift, so they know they are being

11

well looked after. The fondant is a block, or brick, of hardened sugar.

Friday, 8th January Well, the deed is done. On my 50th birthday, on Wednesday, at eleven in the morning, in the church of St.Mary the Virgin, Friston, I married Fiona Hudson, once my landlady in Baker Street, and now my dear wife. It was cold, but bright all day. The ladies looked very fine in their dresses and astonishing hats, and the bride outshone them all. Fiona's brother gave her away, and John Watson was my best man, of course. We were a small party in the church – some twenty of us – including young Wiggins, erstwhile captain of my Baker Street Irregulars, now a giant of a man, and a captain of meat porters at Smithfield, with his family; Inspectors Gregson and Hopkins of Scotland Yard, and their wives, and Mr. and Mrs. Rudyard Kipling, come over from Burwash. We lunched at home. The Inspectors had brought with them a charming wedding cake, which Fiona and I cut with my old swordstick blade, to applause.

During the afternoon I took photographs of the company in the garden with a "Box Brownie" camera – made in Rochester, NY, I now see on the label – a gift from H.H. Shri Sir Mansinghji Soorsinghji, the 25th Thakore Sahib of Palitana, in Gujurat. (He consulted me in 1890, and I was able to help him. The case was not suitable for Watson's records.) Unfortunately, because of my lack of skill with the camera, none of the pictures came out successfully.

In the evening there was a party, at which we were joined by several neighbours and their families. My birthday present from Scotland Yard was a phonograph machine, with wax cylinders that play various polkas – the Berlin Polka, and the Alsatian Polka, are two – to which the young ones danced madly round the drawing room. Jas. Hiams gave me a stuffed badger. The Kiplings brought me an English translation of Maeterlinck's *Life of the Bee*, and my brother

12

Mycroft sent me a painting by one of my favourite painters, the Georgian Niko Pirosmani, showing a traditional wedding of that country. Watson and his wife, Rekha, gave me pair of grey slippers from her country, lined with red velvet, decorated with pearls, and with exotically turned-up toes. If I wear these I shall look like a wizard, I said. You *are* a wizard, Holmes, said Kipling, which pleased me greatly.

But to complete the joy of the day, news arrived that MCC had won their match against the State of Victoria, in Melbourne, by 185 runs. Victoria needed 297 to win, but only got 111, despite an excellent innings by Trumper.

We had asked that there be no wedding presents from our guests and well-wishers. I gave Fiona a ring of three old European round-cut diamonds, and she gave me the writing desk in South American rosewood at which I sit at this moment. Soorsinghji, characteristically, ignored the ban on wedding presents and sent Fiona a chintz jacket and neckerchief with a glazed printed calico petticoat, all dating from the1770s. I have also broken the rules, by getting Fiona a second present. It is a Gresham piano in walnut, which will arrive tomorrow as a surprise. Weekly sessions with Mlle. Zabiellska from Eastbourne will bring her back to the high standard she had before she came to London. I could not countenance a piano in Baker Street, but now I am a different man, and I have made amends.

On leaving us, Gregson asked me how I will survive without crime. It is your lifeblood, Mr. Holmes, he said. I tried for a witty reply, but could not find one. I do not know, I said, but I am very happy just now.

Saturday, 16th January A most terrific gale in the Channel on Wednesday night. The large steamer, *Ranza*, 5272 tons, belonging to the Caledonian Steamship Company, on the Thursday narrowly escaped being driven ashore under the eastern cliffs at Seaford. She was outside the port of Newhaven waiting for coal when she was caught in the gale between one and two in the morning, and breaking away from her anchor, she began to drift in towards the East Cliff. She sent up distress signals and the Newhaven harbour tug went out and managed to get her head to sea.

Trenched, dug and manured the kitchen garden area with Bill Frusher, and sowed cauliflower, cabbage, spinach and lettuce, and a few potatoes in frames.

A late wedding present has come from an old friend in Berlin, Konstantin Eugenides, one of the greatest burglars-by-stealth in Europe – an antique sapphire and diamond brooch for Fiona, and a pair of engraved cuff-links in old red gold for me. I cannot help but wonder whose they were.

Tuesday, 19th January We have had a spell of severe frost, and in consequence the bees have been confined to their hives. Sixteen degrees of frost is too low for their comfort.

Monday, 1st February Right on to the end of January we have had practically no snow, yet the weather has been unsettled, with rough winds, keen frost, and a variable temperature. Bees have been much confined and had few thoroughly good flights. Many dead bees have been carried out of the hives by the workers whose task that is.

Pruned our fruit trees, and Fiona, with my assistance, planted delphiniums, peonies, anemones, and three roses – *Blanche double de Coubert, Cécile Brünner* and *Général Kléber*.

Thursday, 18th February The fields have been transformed as if by the hand of an enchanter, and today the landscape is one unbroken

14

expanse of dazzling, stainless snow. Downy flakes have fallen silently in the night, and each hitherto bare bough sustains a heaped-up pile of one or two inches. No breath of air disturbs the serenity of the scene. I made sure this morning that the lower entrances to the hives were clear of snow and debris. There is an upper entrance, which is left open in Winter to diffuse moisture from the hive. I looked in there and saw the faithful guard bees resting just inside. I also put my ear to the hive and at first heard nothing. I have an old stethoscope of Watson's - how I acquired it, I cannot for the life of me remember - and applied it to the hives. I then heard the very faintest humming sound. I asked Jas. Hiams about this and he told me the bees keep warm in very cold weather by moving their wing muscles in a kind of shiver.

Monday, 29th February 32F, with snow still on the ground, but wet and heavy. The back of Winter's deep chill is broken. Spring will find a way. The bodies of dead bees are still being brought out.

Friday, 4th March I am about to begin two days of struggling up to the attic with my books and papers, helped by young Tom from the Post Office. Two days of sawing, hammering and banging by Eric the carpenter have put up the shelves in my study and in the hallway outside, so my stuff is no longer cluttering up the house.

Saturday, 12th March 45F this morning - a cool day. I saw a bumblebee on the path in my front garden. She was moving slowly and looked cold. I lowered a finger and she clung on. I hoped my body heat would transfer some warmth to her. I had my magnifying glass with me, and it was a great pleasure to have some minutes to examine her closely. Could it be that that extraordinarily thick coat allows bumblebees to fly in temperatures when honey bees cannot? I noticed how very long her tongue is, and that it extends even further with a light, feather-like flexible tip. This bee can clearly reach into flowers that no other insect can, and, I imagine, has the size and strength to open flowers that snap shut, such as snap dragons. I took her into the house and gave her a drop of honey, which she at once began to lunch on, before flying away without a word of thanks.

Wednesday, 16th March While out walking yesterday afternoon on Pevensey marshes we saw a wonderful sight – many hares, half dancing, half boxing, in the fields. I am not ashamed to admit it, but I have no explanation – my eyes filled with tears.

Friday, 18th March A lovely spring day – sunny, windy, quite cold. On today's visit to the hives there were piles of dead bees outside each hive, casualties of the Winter brought out by their sisters. I swept them away, while the sound of Fiona practising Scarlatti with Mlle. Zabiellska came to me through the open window.

Saturday, 26th March This late, almost flowerless, Spring points to the need for a supply of pea-flour for a few days till pollen can be gathered. Jas. Hiams tells me the way a bee-keeper knows Spring has arrived is when the bees are seen returning with pollen on their legs. This shows the queen has started laying.

Bill Frusher and I pruned our hardy roses today.

Saw Dr. Wardleworth yesterday. He listened to my chest, looked at me quizzically and asked if I might be able to give up tobacco altogether, now the stresses of London life are past. I told him Stirzaker's No.5 shag is as essential to me as oxygen, and I have done enough for my health in giving up cigars and cigarettes. But now I have both the good doctor and my wife ranged against me. She particularly disapproves of my having a breakfast pipe of the plugs and dottles of the previous day, and taking a glowing coal from the fire to light my pipe. I have told her that when one marries a bachelor of confirmed habits, one cannot change him, as in a Grimm's fairy tale, overnight.

Sunday, 3rd April Jas. Hiams and I inspected my bees this morning. This is the moment when (as the poet Shelley has it, addressing the west wind):

> *Thine azure sister of the Spring shall blow*
> *Her clarion o'er the dreaming earth, and fill*
> *(Driving sweet buds like flocks to feed in air)*
> *With living hues and odours plain and hill;*

Scenting the freshness of the air, the bees seem to revel in it. They move about the entrance to the hive, examine the doors and porch, meet and salute each other, and fly for a moment in front of the hive. With each succeeding sun, the bees move abroad in larger numbers. The queen communicates to all around her that the hour has come for which they have lived and waited through the long months of Winter. She passes from cell to cell, examining each and depositing in it a tiny egg. Nurse bees then lavish their care on it. During three days they will hatch it, and then when the grub appears, they feed it with honey and pollen stored in the adjacent combs. Then they seal the cell until the grub, now transformed into a nymph, emerges a week later as a perfect bee, to share the labours and to participate in the busy and often hazardous enterprises of the colony.

Other bees are cleaning the hive, sweeping the floor-board clear of fragments of broken comb, pollen pellets and dust, while their sisters on guard duty patrol the entrance to the hive to prevent the entrance of robbers.

Bees collect four things: pollen for protein, nectar for energy and conversion into honey, water, and propolis (from leaf buds) to use as glue in the internal construction of the hive. Where nectar is, they sip it; where there is pollen, their feathery hairs collect it, and they bear it home in the little baskets on their hindmost legs to feed their grubs. Propolis is a sticky, resinous substance found in pine, horse chestnut and other tree buds, used by the bees to fasten joints in the furniture of the hive, to exclude draughts and to keep the hive watertight. In the wild, those bees whose task it is to supply material for the building of the combs have fed themselves from stores of honey and now, clinging to each other, hang motionless during many hours until clear scales of wax appear on their abdominal plates. They hand these scales over to the builders who, taking them in their mandibles, construct the comb whose elegance and strength teaches such a profound lesson to human engineers. In Langstroth hives, like mine, there are frames of wax foundation already sup-plied, which the bees draw out into cells.

Planted two new Cox's Orange Pippin trees.

Monday, 4th April Easter Day was cold, dull and cheerless, inter-spersed with driving snowstorms. I am still supplying artificial pol-len and feeding the bees with thick syrup, while providing a plentiful supply of water. If I can tide over the present untoward spell of bad weather, the promise all around of bee-forage ready to burst into bloom will sustain my hopes for better things ahead.

Friday, 8th April Weather conditions remain bad. Cold NW winds, interspersed with hail, sleet, rain and thunder; and what of the poor bees during such a long spell of inclement weather? A glimpse of sunshine brings them out, and then the cruel storm beats them down in hundreds, never to rise again.

Heard the cuckoo for the first time since I was a boy. As when I saw the dancing hares, my eyes filled with tears.

Sunday, 10th April Hung my collection of portraits of notorious criminals in my study, among them Diogo Alves, Mary Bateman, William Henry Bury, Pierre François Lacenaire, Thomas Cream, Hadj Mohammed Mesfewi and Sophie Charlotte Ursinus. And Moriarty, of course.

Monday, 11th April A lovely sight this morning – large lumps of gold on hundreds of workers' legs as they arrived back at the hives.

Sowed parsley, spinach and radishes.

Thursday, 5th May A bright, windy day, the soul of Spring! How can I have gone so many years without allowing myself, surely the sharpest-eyed of God's human creatures, to notice such things? I opened the front door this morning and a sudden gust of wind blew a crowd of pink blossoms in through the doorway and along the hall. No-one was about; I did a little dance of joy.

At dinner last night our neighbour and fellow apiarist, Herr Rainer Tegetmeier, told us that his bees – all Blacks – have been coming back with a yellow stripe formed in a rectangle on their thorax. He thought at first it was a sign of disease, and consulted the *British Bee Journal*. The experts there told him that some flowers are set up with triggers that slap or spray a puff of pollen on the bees' backs as they enter or leave. Pollen on their backs is almost impossible for the bees to remove, so it seems those flowers are very clever. They know how to brush their pollen onto the very spot where it is most likely to get transferred into other flowers of the same species.

Saturday, 14th May The long-delayed Spring came in with May Day, and the cold, inclement winter weather has given way in the last fortnight to summer heat. The bright yellow of the dandelion shows plentifully in the meadows, while the anemone and wild cherry are blossoming in abundance in the woods.

We had tea in the garden, down by the apiary. We watched a worker, home at last, walk slowly across the alighting board loaded with pollen. We saw two colours of pollen while we looked – a bright yellow and a soft white.

(I realise I have not told my readers, some of whom may be bee-keepers, that I follow the normal practice and keep a daily diary about my bees. It is a log where I record technicalities such as the daily temperature, how many frames the brood is on, whether I have been able to see the queen, if queen cells have appeared, what the mood of the bees is, and so on. In this journal, however, which may one day interest a few readers as well as its author, I am sure a mountain of such detailed information would be unwelcome, so I give here only selected extracts from those diaries.)

Thinned seedlings under Bill Frusher's expert eye.

Saturday, 28th May Dull weather, with some rain and a little sunshine, is my weather report for the last fortnight, and its adverse influence has retarded breeding. I am still feeding my stocks to prevent the bees from starving, so that even with a change for the better we can hardly hope to see any honey stored for another two weeks.

Last week sowed cinerarias, primulas and wallflowers, and planted out what I hope will be a splendid wall of hollyhocks.

Friday, 3rd June It seems that all over the country this has been a disastrous season for apiculture. All my correspondents agree that the circumstances are abnormal and almost unprecedented. I myself have many dead bees, and many more weaklings. But the closing days of May were gloriously bright and sunny, so we may yet see a marvellous change.

The hoe is in almost daily use to keep beds and borders free from weeds. Planted out tomatoes and cucumbers.

On Jas. Hiams's recommendation, I installed a robbing table about fifty feet from my hives. This is a box about three feet high on which

I can put sticky combs, to distract robber bees from the hives. The robbers are not about yet – they will come later in the Summer when honey stocks are at their most plentiful, which the robbers can smell – but it is good to be ready.

Thursday, 16th June After twelve hours' rain yesterday the sun is shining this morning, and on my return from my sea bathe I saw bees busy on my neighbour's rhododendrons. I had thought that the rhododendron was not a plant for bee forage because of the toxins in all parts of the plant, including the nectar. It is possible that the toxins decay over time, so that the level of toxicity is negligible by the time the honey is eaten.

A propos forage plants, I was very nearly involved in a fight this evening. I was sitting in the low-ceilinged Snug in the Tiger Inn, with Jas. Hiams and two other gentlemen, not from this village. We were talking of bees when one of the gentlemen asserted that the buttercups so prevalent in our pastures are of little use as bee plants. Buttercups, he said, with the air of one who is very rarely wrong in matters of this kind, are, in general, unpalatable plants, owing to the presence of an acrid poisonous principle in them, which can cause poisoning in livestock. Nonsense, said Jas. Hiams. My bees have always visited the buttercups. You may find it of interest to know, said the gentleman, that in recent years the pollen of buttercups has been proved to be actually injurious to bees in Switzerland, and elsewhere in Europe. But not here, said Jas. Hiams. This is Sussex, not Switzerland. In either place, said the gentleman, the harmful nature of buttercup pollen is the reason why the flowers of the species are completely neglected by hive bees. Their instinct warns them to leave the flowers alone. I rose and tried to intervene, but was cut off after my first two words. What you know about bees, with respect, Mr. Holmes, said Jas. Hiams, could be written on a bee's backside. At this atrocious thrust I sat down. I just told you, sir, he said to the gentleman, rising, I just told you, my bees *do* visit

21

the buttercups, they have done for years, and *there is nothing wrong with that pollen.* That cannot be, said the gentleman, calmly. You are mistaken. Cannot be! roared Jas. Hiams. Mistaken! I'll bloody well show you a bloody bee in a bloody buttercup – and suddenly his hands were gripping the gentleman's lapels. Our landlord, Thomas Birtwistle, then appeared at our table. He stands six foot seven, and weighs sixteen stones. Jas. Hiams sat down, but left in a rage immediately afterwards, with half his pint of good Harvey's undrunk.

Fiona now has a bicycle, bought for 19/- from Victoria, the wife of the Rev. H. R. Cadwalladr, who officiated at our wedding. It is a drop-frame Columbia "Ladies Safety" model, and she intends, with my full approval, to ride it through the lanes in rational dress, that is, wearing the eminently sensible "bloomers", with veil and an over-skirt.

Wednesday, 6th July As I sit in my garden on this warm afternoon, and observe the flight of my bees, I feel again a desire I have always felt, to know how the world appears to other beings. On this question our knowledge is still extremely defective. Have bees, for instance, the same senses as ours, or fewer, or more? Ultra-violet rays which are invisible to us are visible to some ants and crustacea, and may well be seen by bees. Why should we assume that there can only be five senses? Sound is the sensation produced by vibrations of the air striking on the drum of the ear; when they are few the sound is deep; as they increase in number it becomes more and more shrill, and finally it ceases to be audible to humans. Light is the effect produced on us when waves of light strike the eye. When 400 millions of vibrations of light strike the retina in a second, they produce the sensation we call red, and as the number increases the colour passes into the orange, then yellow, green, blue and violet. Below 400 millions, we have no organ of sense capable of receiving the impression. The familiar world of sense that surrounds us may be a quite different place to some animals; to them it may be full of music which we cannot hear, of colour we cannot see, of sensations we cannot feel. Here is a wide, and as yet almost untrodden, field of study. Charles Darwin once said that the brain of an ant was the most marvellous atom of matter in the world. Surely that of the bee can be little less so.

Fruit trees covered with netting, and winter greens planted.

Monday, 18th July A letter today from France. Since 1900 I have been in occasional correspondence with M. Paul Ulrich Villard, of the École Normale Supérieure in the rue d'Ulm, in Paris. He was studying the radioactive emissions of radium when he identified, in that year, a new type of radiation that he first thought consisted of particles similar to known alpha and beta particles, but with the power of being far more penetrating than either. We were joined in our correspondence in 1902 by Herr Professor Max Planck, who thrilled both M. Villard and myself by asserting in one of his first letters that the energy of oscillators in a black body is, to use his ugly word, "quantised", a phenomenon that classical physics is unable to express. At the time he thought little of this "mathematical trick", as he called it, and I had to argue for some time with him, relying heavily on Boltzmann's statistical interpretation of the second law of thermodynamics, before he could reluctantly accept that this "trick" actually constituted a fundamental change – a revolutionary change – in our understanding of how the world works. At that point our letter-writing circle was expanded by the inclusion of three boisterous young men, all in their early twenties, who were working in Bern, in Switzerland: Herr Conrad Habicht, a teacher of mathematics; Herr Maurice Solovine, a student of philosophy; and Herr Albert Einstein, who is a patent clerk. These are the minds, I am sure, that will bear this new thinking into the twentieth century, following the lead that Herr Planck has provided. His is an astonishing mind, whose genius is only matched by his modesty.

Received today three healthy bushes of Russian Sage, the gift of Mr. William Robinson, the eminent Irish gardener who has Gravetye Manor near here, in East Grinstead. He has introduced it to this country. His note says that though the plant is called "Russian", it is not from there. It grows in various forms from Iran to India on open, well-drained ground. Apparently it is a great favourite with the bees. It has spires of purple flowers and finely-shaped grey-green foliage.

Astonishing weather – storms and hail, high winds, thunder and lightning. All four hives stayed upright.

Sunday, 16th August The last two days have been glorious. An eight-hour day was scorned by the bees; even ten hours was too short for some of them. The rush and hustle were wonderful to see, and the constant stream of bees, outgoing and incoming, required the full stretch of entrance at the front of my hives. These two days glutted the frames and the supers above them.

We have got to know a neighbour in the village, Capt. G. Minty, who has a splendid automobile – a De-Dion Bouton Populaire. My first photograph is of Fiona, pretending to drive it, with Minty hovering slightly anxiously behind.

Wednesday, 19th August Orientation flights seem to take place at about three o'clock in the afternoon in my apiary. As the sun moves across the garden and lights on a hive, the bees come out. They fly back and forth repeatedly, starting low over the hive and then moving higher.

Thinned out dahlias, and much gathering of fruit. Jas. Hiams says soft fruit gathered early in the morning keeps better.

Thursday, 20th August I played my violin up on the Downs at dawn, all alone and no doubt, from a distance, looking like a madman.

Saturday, 22nd August Called this morning to help with a wild bees' nest in an empty house down the lane. The bees had entered from a small hole at the bottom corner of a second-storey window. William, a young local man, put on full protective clothing, including a veil, went up a ladder and started pulling off the boards over the window. He then reached into the room with a long knife, cut the golden four-foot combs and laid the pieces in a cardboard box that I, equally well protected, held up to him. When he was done we took the combs and put them in a hive as best we could, sweeping the bees in after them.

Tuesday, 25th August This morning after breakfast I read to Fiona the entry for Edward Agar in my criminal scrapbook. Agar was a highly intelligent man who by the age of 39 had accumulated a fortune on the stock market, and lived in respectability and comfort in Shepherd's Bush with his mistress, Fanny Kay, a barmaid. Agar learned from a man called Pierce, who had worked for the South Eastern Railway, that gold ingots were regularly sent by train from London to the Banque de France in Paris. Agar decided to steal the gold simply for the daring of it. With the help of a guard on the S.E.R. called Burgess, he managed make copies of the two keys needed to open the new Chubb safe installed on the bullion train. He and Pierce boarded the train at London Bridge on the evening of May 15th, 1885, and Burgess let them into the guard's van. Before the train reached Folkestone the two men had transferred nearly two hundredweight of ingots to their carpet bags. Agar's admirable attention to detail, and forethought, was such that he had brought with him lead weights which were substituted for the gold, so the caskets would weigh correctly on arrival at Boulogne. Amusingly, when the theft was discovered, the English police insisted that it must have happened in France, because their own security was beyond reproach.

Unfortunately for Agar he was arrested not long afterwards on a completely unrelated charge of financial fraud. He was found guilty and sentenced to transportation. He asked Pierce to make sure that Fanny Kay received his, Agar's, share of the gold. Pierce, however, wanted to keep it all for himself, and when Agar learned of this in Pentonville he "sang like a canary", as they say in the force, and confessed to the robbery. Pierce and Burgess were duly convicted on his evidence. But then, in a sensational ruling, the judge ordered that Fanny was entitled keep certain bonds which had been bought with Agar's share of the gold. She became rich, and the brilliant criminal went off to Australia. I have not been able to trace his subsequent career, or hers.

Thursday, 1st September Planted bulbs of narcissus, hyacinth, tulip, crocus and snowdrops. Turnips stored, and much watering of the asparagus beds.

Wednesday, 14th September This afternoon, it being bitterly cold and rainy outside, I was going through the first of my scrapbooks, and came across the case of Adelaide Bartlett. She was tried at the Old Bailey in April 1886, the charge being that she had poisoned her husband, Edwin, with liquid chloroform. (For readers of this journal unfamiliar with chemistry, chloroform is an organic compound, one of four chloromethanes. Trichloromethane was synthesised independently by two groups in 1831; Liebig carried out the alkaline cleavage of chloral, while Soubeirain obtained the compound by the action of chlorine bleach on both ethanol and acetone. In 1835, Dumas prepared the substance by the alkaline cleavage of trichloroacetic acid, and Regnault later prepared it by chlorination of monochloromethane. By the 1850s, chloroform was being produced on a commercial basis using the Liebig procedure, and was widely available to the public.) Adelaide had married Edwin, a wealthy grocer, when she was nineteen; ten years later she wearied of Edwin and began a relationship with a Wesleyan minister called Dyson. At this time Edwin made a will, leaving everything to Adelaide, and soon afterwards became ill with a gastric complaint, from which he died in January 1886. When massive amounts of chloroform were found in his body at the autopsy, Adelaide was charged with his murder. But the prosecution could not show how she could have administered it to him without his being aware, because of the burning sensation chloroform gives when ingested, and there were no traces of chloroform in his windpipe. She was acquitted. Sir James Paget, consultant surgeon at Barts Hospital, was involved in the case. After the trial, he said, "Now she has been quite properly acquitted, she should tell us, in the interests of science, how she did it."

Tuesday, 4th October Storing of plentiful harvest of apples and pears. Lifted late potatoes. Planted box, lavender and camellia in the garden at the front of the house.

Blackberry and ivy still offer the bees sustenance as the days grow colder, but foraging becomes an ever more precarious task. Rain and high winds claim their victims. The queen ceases to lay, and the bees collect in clusters on the central combs. There they will hang together until Spring returns.

Wednesday, 13th October The weather is changing, and the bees feel it. Summer breezes are now cold draughts. If there is a crack or hole in the hive, wind and water will find their way in. The bees are busy sealing these holes with propolis, scavenged from wherever they can find it. They have given us a good honey harvest, and are now preparing to rest.

Saturday, 12th November This morning stored turnips, beetroot and artichokes. Staked young trees and mulched, and pruned apple, pears and plum trees.

Saturday, 19th November I have been raking leaves, draining and storing the two garden hoses, emptying the gutters of leaves and putting my garden paraphernalia away. There is a feeling of early snow in the air. My bees are tucked up with their warm wraps on.

Only four stings, all season!

Friday, 9th December I have some interest in Tibet, having visited there in the early '90s. I have been reading with disgust an account of the military adventure recently conducted there by Major Francis Younghusband, acting under orders from Lord Curzon. Younghusband was supposed to be doing nothing more than settling a dispute over the Sikkim-Tibet border, but when he was about one hundred miles into Tibet, en route to Lhasa, his expedition massacred many hundreds – perhaps thousands – of Tibetan monks at a place called Guru. The Dalai Lama, Thubten Gyatso, whom I consider a friend even though we have not corresponded for eight years, had to flee to Urga, in the north of Mongolia. Younghusband has received the Order of Knight Commander of the Indian Empire for his gallant action.

With Bill Frusher's help, prepared a mushroom bed.

Friday, 16th December This morning I was walking in deep snow in The Pines, originally a gravel pit, and now a meadow. Suddenly my left leg broke through the snow into watery ground below. While I was pulling on that leg it came out, but without my boot, and immediately my right leg went down up to the knee, and I fell backwards. I rested for a moment on the snow and considered my position. Eventually I sacrificed the left boot and limped home, wet and muddy, to the amusement of my wife.

Friday, 23rd December Carol singing round the village with some dozen neighbours, led by the Rev. Alfred Evans and ending up in the Tiger. I bore the lantern on a pole.

There was a traditional play acted in one of the Tiger's back rooms later. The characters, in full costume, were a countrywoman called Molly; Father Christmas; St. George, the Turkish Knight; a Prince of Morocco; a Doctor, and a boy called Johnnie Jack. It was full of bombast and strange songs, and I could not make head or tail of it.

Sunday, 25th December A very mild and bright Christmas Day, the bees flying strongly at 9a.m. This they continued to do merrily until 2p.m., when the wind became rough and cold, and the bees took shelter from it in the hives. We have had high winds of late, but no damage has been done to my little apiary. All four colonies are strong so far, with stores plentiful, although rapidly diminishing, so, in order to keep all going right, four cakes of candy will not go amiss, and they shall have it.

For lunch we shall have the Whitakers of Gore Farmhouse, the Bushells from the Old Parsonage and the Hadleys of Filching Manor. In a moment I shall wrap Fiona's present – an acrostic ring from Cartier, with seven stones arranged so that the first letter in the name of each stone spells out the word "dearest" – diamond, emerald, amethyst, ruby, emerald, sapphire, and topaz.

Wednesday, 28th December To The Duke of York's last night for the opening night of a charming play, *Peter Pan*, by Mr. J. M. Barrie. We went backstage at the end and were introduced to Gerald du Maurier, who plays both Mr. Darling and the abominable Captain Hook, and to Miss Nina Boucicault, who plays Peter.

Saturday, 31st December The last day of the year, and preparing to go over to our friends the Horsnails for the customary celebrations. The roads are clear, and we shall shortly be borne there and back in Minty's DeDion-Bouton.

I give here some general advice on the handling of bees, gained over the last twelve months, and 24 cures for stings, all of them infallible, gathered from Jas. Hiams, William Tompsett and other bee-keepers on the Weald.

In manipulating, never stand in front of the hive. You interfere with the flight of the bees if you do so.
Never manipulate during rain and stormy weather.
Work quietly, and avoid all haste and nervous movement.
Do not breathe on the bees, as the breath and perspiration of man and animals are repugnant to them.
Select a time when most of the fieldworker bees are absent from home, as they are more irritable than young bees.
Avoid crushing a bee. The odour produced excites the colony.
Use smoke moderately; too much does harm.
Always wear a veil, but avoid gloves if you can bring yourself to do so.
Wash your hands before every operation.
Use different gloves and tools for each hive, to avoid transferring infection, or rinse the tools in soda water between hives.
If you are stung, remove the sting immediately, by sliding a knife across the skin, because the sting continues to pump venom by itself.

Cures for stings: laudanum; olive oil; vinegar; dock leaf; urine; rubbing with an onion; saliva; common salt; the juice of a plantain; tobacco; whisky (rubbed in, of course, not drunk, though drinking it may be of some help in the long run); chalk dust; ammonia; baking soda; honey; black earth, or clay; linseed oil; butter; hog's lard; soap; sage leaf; kerosene; bruised parsley.

(A correspondent, M. Bourgeois, of Tunis, sends me the following additional recipe for the prevention and relief of stings: menthol, 30 grammes; alcohol, 40 grammes; glycerine, 100 grammes. Well mix the ingredients and apply to the face and hands.)

One soon comes to recognise the difference between the genial hum of a friendly bee and the angry buzz of one on the warpath, and the bee-keeper must know the liberties a bee will resent.

And finally, to close this momentous year, I am delighted to record that the good Watson has done something most unexpected, and most charming. He has sent me a photographic studio portrait of himself in a silver frame, with the request that I keep it on the table in my study as a memento of our days together in Baker Street. I am touched. It is before me as I write, and I shall certainly include the photograph in this journal if it is ever published.

1905

Thursday 5th January In 1851 the magnificently-named Reverend Lorenzo Lorraine Langstroth – apiarist, clergyman and principal of a college for young ladies in Philadelphia, Pennsylvania – built on Johann Dzierzon's observations that the spaces left by the bees between their combs were about one inch wide, and proposed that if a "bee space" of this size were left between all separable parts of the hive, the bees would leave them free of their building materials, such as beeswax and propolis, thus allowing the frames to be removed easily and without endangering the bees. Langstroth made a prototype, and hives on the Langstroth principles and dimensions are now almost universally recommended, though I have examined a Dadant hive, which is more common in Europe than the Langstroth, and which has some advantages over it for extracted honey production. Supplies for Langstroth hives can easily be secured from America, and increasingly so in this country. I use the ten-frame size.

The Langstroth frame is long and shallow – approximately 18 inches long by 9½ inches deep. The shallow frame permits the use of a low, flat hive that can be easily tiered up one, two, three and four storeys high. This is an advantage when one is seeking extracted honey, for all one has to do when the bees need more room is to add upper storeys as fast as the bees require them, and then at the end of the season extract at leisure.

I must add that both the pioneering apiarists mentioned above seem to have been extraordinarily unlucky. Have the bees had a kind of revenge on them for their cleverness in exploiting the bees' labours? Langstroth could never patent his hive, despite tireless efforts. It was copied everywhere, and he missed the chance to become a millionaire. Dzierzon, a Polish priest of the Roman church, discovered some sixty years ago that drones are produced parthogenetically – that is, without fertilisation, or, as the common expression is, by virgin birth. He published his findings in the Bavarian gardeners' magazine *Frauendorfer Blätter* between 1842 and 1844 and was condemned by the church authorities for doing so, though why they could not stomach this idea, when their religion is based on it, I cannot conjecture. This led Dzierzon to question the doctrine of papal infallibility, which had been promulgated not long before. He had demonstrated the matter beyond doubt, and if the Pope denied it, how could the Pope be infallible? For this impertinence Dzierzon was stripped of his priesthood and ex-communicated. I salute this latter-day Galileo of the bee world. He is a hero of patient and accurate observation, rational thought, and courage.

As I re-read in my scrapbook this evening the case of Madeleine Smith, I realise that I have lost count of the murder cases known to me where thwarted love plays a major part. Fifty years ago she was tried for poisoning her lover, Pierre Emile l'Angelier. Because her father was wealthy, and a leading figure in Glasgow society, she was forbidden to see l'Angelier, who was a humble packing clerk. She was only twenty, and wrote the young man many passionate letters. When she at last decided it was hopeless, and told him that she could write to him no more, l'Angelier threatened to show her father the letters unless she met with him and continued the relationship. Some time later he became ill, suffered convulsions, and died. The autopsy showed his stomach was full of arsenic, and Smith was arrested and charged with his murder. Did she poison him by putting arsenic in a chocolate drink? Or did he take his own life, in despair? The jury returned the uniquely Scottish verdict, Not

Proven, which seems to me to signify, "We have some doubts on the evidence for her guilt, but we suspect she did it," and Smith, aged 22, walked free.

Friday, 6th January We have now exceptionally large quantities of hazel catkins, but the pollen from them is being wasted by the prevalent gales and heavy rains, which keep us indoors. The song of the mistle thrush is increasing in vigour, but this is said by Mrs. Hiams, with sighs, to betoken continued wet weather.

Monday, 9th January Taking advantage of a recent fine day – one of only two or three we have had for some time – I made a hasty examination of my stocks of bees. They were badly in want of food, but by equalising the stores, and putting candy over the feed-hole, I have made them safe for at least six weeks to come, when – surely – I shall have suitable weather to make a more careful examination.

For my birthday last week Kipling gave me a fine edition of *Gargantua and Pantagruel,* by François Rabelais, illustrated by a Mr. William Heath Robinson, and signed by him. I do not know the book, but it is certainly a big one, and I look forward to reading it to while away some of the hours of foul weather we are having.

I also received a consignment of tinned pineapples and tinned bananas from the State of Queensland, Australia. It was from Kees de Jong, once master of the *Friesland,* a Dutch vessel that went down in the Gulf of Carpentaria in '95. Working only from London, I was able to clear his name of the charges against him of drunkenness and incapacity, and point the finger of justice at the Dutch freight company and the English insurance company who, working through a corrupt Chief Engineer, had planned the sinking of the steamer, regardless of the consequent loss of life of more than half the crew.

Tuesday, 17th January With Bill Frusher laid new gravel on the paths, and covered the seakale and rhubarb to blanch.

Sunday, 29th January It has just started to snow. This is a bad time of year for most bee-keepers. We do not know what is going on in the hives, and it is no use opening up the hive to look, as that would not be good for the clustered colony. In any case there is nothing one can do to help them, other than feeding. So I check to see if the weight of the hive is light, indicating a lack of food in the box. One can feed them without opening the hive, by lifting the cover board and putting the food on top of the combs directly over the nest.

Wednesday, 1st February Another filthy afternoon, with high winds and driving rain off the sea. I am reviewing, among others, the case of Harry Benson. It is a case which reveals some of the widespread corruption in the Metropolitan Police in the 1870s, corruption which, I am glad to say, had largely been eradicated by the time I began my association with some of its detectives. Benson, whom I believe to be still alive and living in New York, was – and perhaps still is – a consummate swindler whose early speciality was to pose as a charitable man raising funds for war refugees. With a Scot called Kurr, he bribed a senior detective, John Meiklejohn, to obstruct police investigations into horse-racing frauds. He then began blackmailing Chief Inspector George Clarke and that once brilliant detective, Nathaniel Druscovich, who was made Inspector at the astonishingly young age of twenty seven. Both men had financial difficulties, and had accepted loans from Benson. Thousands of pounds passed to and fro. In September 1876 Benson changed his speciality, and swindled the Comtesse de Goncourt out of £10,000 by persuading her to back non-existent horses. Meiklejohn and Clarke did their best to disrupt the police investigation, but Benson panicked and fled to Holland, and was arrested there. He received fifteen years penal servitude, and the three officers ten years each.

Exactly the kind of case I should have been keen to take an interest in, but at that time I was, alas, too young to be known to the force.

Meiklejohn and Druscovich were shown to have been involved in a previous duplicity, in 1873, when they allowed the fraudster, Austin Bidwell, and his brother, George, to escape to the United States after they had duped the Bank of England with bills of exchange forged by that non-pareil of counterfeit engravers, George Macdonnell. The sums involved were in excess of £100,000. The Bidwells were brought back from the States and both sentenced to penal servitude for life, as was Macdonnell, but the Bidwells were released on grounds of ill-health about twelve years ago. The presiding judge, Mr. Justice Archibald, created a precedent – he sat throughout the case with a gun concealed in his robes, because it was thought that there might be an armed attack on him in court. Macdonnell made a rambling plea to the court which went on for an hour. It contains this touching sentence: "It has become, as one of the newspapers said, an art – I mean, fraud of this description – and, although a very wretched, unhappy, miserable, and contemptible art, it may be, to a certain extent, called an art nevertheless." I have two examples of his sublime handiwork in my scrapbook, and still cannot tell them from genuine banknotes, even when I look at them with my glass.

Saturday, 4th February The month of February has opened with a few mild sunny days during which my bees have been able to take free flights, arrange their food supply – which Jas. Hiams tells me is probably winter-flowering honeysuckle, snowdrops and winter aconite – and in many cases no doubt breeding has started, too. I judge this to be so by the number of bees at the watering-plates, and I have had to give a fresh supply of water each day.

I read last night, while a violent south-westerly gale rattled our window frames, an interesting account of bee-keeping in Palestine. It was in the *Bulletin de la Société d'Apiculture de la Somme*. Nectar is apparently abundant in Palestine. The first harvest, and the most important, is in April, at the time of the flowering of the orange trees; the second is in May, from mountain flowers, and lavender and rosemary; and the third in June, from thyme.

I am working on a rendering of J. S. Bach's *Goldberg Variations* for violin and viola, though where in east Sussex I shall find a viola player to accompany me, I do not know.

Saturday, 10th February Spring cannot come too soon for me. It is below 32F this morning. Frost on the window, and the grass covered in hoar frost. It is the coldest day we have had this year and we are now in the second week in February. It shows how far from normal the weather is this Winter. I wonder how this will affect the bees.

Tuesday, 13th February With Bill Frusher manured the strawberry plantation and the gooseberry bushes, and pruned the peach and nectarine trees.

Thursday, 23rd February A week of mild weather has invigorated the bees with new life, while the sunshine has opened the crocus-blooms a month earlier than usual. The sheets on the washing line are blowing gently in the breeze as Pearl Thomas pegs out the washing in the garden below me.

Sunday, 26th February Drones have ever been a mystery, and their existence in such numbers has puzzled many wise heads in all times. In a community where the teeming thousands are distinguished above all other insects for their industry, why should a class of idlers exist, gorging themselves on the choicest sweets, and spending their time in luxury? And why, when they have once been called into existence, should the workers so ruthlessly evict them?

It was only about a century ago that the researches of Huber, and other succeeding investigators, convinced even the most sceptical that the only reason for the drones' existence was to fertilise the queen, and, that duty once performed, the far-seeing little workers at once expelled them as encumbrances. Their number is explained by the fact that the queen has to go abroad on her marital flight, and that only after a prolonged journey in the air do the strongest and fittest of her multitude of suitors overtake her. Nature thus secures by an invariable law that the survival of the fittest perpetuates the race. To few has it been given to actually witness that meeting, but Jas. Hiams has, time and again, seen the young queen return from her nuptial flight bearing the unmistakable evidence that mating has taken place during the minutes she has been absent, and that, as a consequence of the act, the selected drones have offered up their existence on the altar of duty.

Thursday, 2nd March It has been very mild and sunny here for the last few days, and bees have been very busy in the aconites, spread like a carpet of gold before them. There is nothing I like better than to see the bees crowding in with their first loads of yellow pollen for the year. I find my queens have started laying.

Saturday, 18th March This month, up to Friday 17th, has sustained its reputation for storms and blustering winds; then came a most pleasant change for the better, and now Springtime has been ushered in with glorious sunshine. The bees have, in consequence, been on the wing in their thousands, visiting every floweret in bloom. The crocus, white arabis, and early blossom of the wallflowers and carnations in our garden were one merry hum.

This morning I had an invitation I simply could not refuse. Looking at me through the window was the sun, shining as though it were a summer's day. I threw down my newspaper, heaved myself out of the chair and went to the apiary. The bees in all the hives put me to shame, flying and bringing in pollen. We took a fine walk from Seaford over the cliffs to Cuckmere Haven, where we saw cormorants, oystercatchers and egrets, then up again and down to Birling, and back round West Hill to home.

Wednesday, 22nd March Bill Frusher has finished digging a small piece of vacant ground beyond the orchard. We sowed beet, beans, early horn carrots and leeks.

Tuesday, 28th March I see that M. Jules Verne has died. I have always felt a warm affection for him, as *Five Weeks in a Balloon* and *Journey to the Centre of the Earth* were a joyful, secret relief to me when I was at school and having to read texts so boring and inconsequential I almost weep remembering them.

Wednesday, 5th April Early April very chilly. The gooseberry crop has, I fear, suffered irreparable injury.

Fortune favoured me in infancy by introducing me into a bee-keeping family. Some of my earliest recollections are naturally entwined around the old-fashioned straw skep and its busy inhabitants. Among the first things I had in mind was an awful dread of the bees' stings, and an intense fondness for their honey. I also remember feeling much pity for the poor labourers whose fate it was to die after having worked so hard all Summer. That feeling, however, soon gave place to joy, when I was supplied with a hunk of bread and a spoon, together with a piece of the delicious honeycomb cut from the skep, making up a veritable feast for myself as the youngest of the family.

A local boy called Ray Kemp got badly stung yesterday. He was close to what was an old brickyard, with several acres of willow-catkins in close proximity, alive with bees foraging. He disturbed them by running under the trees and knocking the branches. When I saw the poor lad's face, his mother had painted it blue, and he was unrecognisable. I gave him a shilling. Ray's father, Jesse, is a wheelwright, and Elsie, his mother, queen of the six big wooden tubs in the village laundry, in Rose Cottage. The boy lives surrounded by stiffly starched skirts and petticoats.

Thursday, 6th April This afternoon I found a scrapbook entry of a case with no forensic aspect, but which is interesting nonetheless. Ernest Boulton spoke with a soprano voice and went by the name of Stella. From an early age he had been compelled by his mother to wait at table dressed as a housemaid. His companion, Frederick Park, known as Fanny, was a solicitor's clerk. Both lived with Lord Arthur Clinton, the M.P. for Newark, who called Stella his wife. Boulton and Park were arrested in April 1870 as they left the Strand Theatre. Bolton was wearing a scarlet dress with white moiré antique petticoats; Park was in green satin with black lace trim. Until only nine years before, the penalty for buggery had been death. It was changed to life imprisonment in 1861, so the young men were in serious danger. The jury acquitted both with cries of "Bravo!" from the public gallery, and a word of censure on the police from the judge, Lord Chief Justice Cockburn.

Sunday, 9th April The weather is starting to improve. Spring is here and Summer will not be far behind. Hefting the hives tells me they are well stocked with food. Already the bees are taking pollen in in large quantities.

Thursday, 13th April On opening the hives I found plenty of stores, just as I expected. The queen has been working hard, with about four frames of brood. Very good. So there we are, eggs, open brood and two of the frames full of sealed brood. There was also a patch of drone brood which was unexpected at this time of year. Then, to my surprise, there were newly started queen cells. Things look a little too advanced for this time of year but there they were, before my eyes. I do not think they are ready to swarm, but they are certainly thinking about it. If the weather stays warm the bees from one or more of my hives could be away in a swarm in something like a week or so. Jas. Hiams will help me prevent that; he recommends removing the new queen cells, and we shall do that tomorrow.

Monday, 24th April I have excelled myself. If only Watson were here to record it. Yesterday I laid a trail of twelve clues for an Easter egg hunt in our house and garden for the Tompsett children and the Hiams's two grandchildren, who are visiting. Each clue was a couplet of verse, which led the children on to the next clue, till eventually they found the hoard of chocolate eggs.

Saturday, 29th April The weather in this part of the world is perfect April – within the space of about ten minutes we have had hail, rain, brilliant sunshine, and then back to hail.

Planted some new alpines in the rockery, and finished pruning the roses.

Friday, 5th May We had a spell of fine weather after Mayday had passed, but the festive day itself came in a perfect roarer. The East Dean children's garland was blown to pieces as they perambulated the village when "going a-Maying". Later on it became calm and bright, and since then the bees have fairly revelled in the sunshine and flowers. It has been a very great help in building up stocks, though the adverse weather in April left a big leeway to make up. The prospects for the season are promising, the sunshine having made a great improvement in the grass crops. Vetches also are growing well, and the more forward fields around us will shortly provide good bee-forage, as will the dandelion and trefoil. Sycamore and horse-chestnuts, too, are bursting into bloom, with many other flowers in the woods and waysides.

I am sitting in the garden writing, and listening to Fiona inside the house practising the Liszt *Hungarian Rhapsody No. 15* with Mlle. Zabiellska.

Wednesday, 10th May I returned from London late last night after spending the day in Percy Circus, Islington, with the Russian Vladimir Ilyich Ulyanov and his wife, Nadezhda Konstantinovna Krupskaya. He uses the *nom de guerre*, Lenin. He is one of 38 delegates to what he grandly calls the Third Congress of the Russian Social-Democratic Labour Party, which seems to meet mostly in the backrooms of public houses. He invited me because *The Hound of the Baskervilles*, Watson's overblown romance that tells the story of a case I was involved in in the late '80s, was published in Russian in 1902. Ulyanov liked it, and wanted to ask me questions about some details of the affair.

As it happened, we had more significant matters to talk about. He told me of the events of January 1905 in St. Petersburg, of which I was only dimly aware. Soldiers fired indiscriminately into a crowd of unarmed strikers and their families who were bearing a petition to the Tsar. He claims 4,000 died that day, and that since then some 15,000 peasants and industrial workers have been shot or hanged, and tens of thousands sent into exile in Siberia. He intends that his party will create an alliance between urban workers and the peasantry to overthrow the Tsarist regime and then establish – I wrote the phrase down – "the provisional revolutionary democratic dictatorship of the proletariat and the peasantry".

He is a man of evident intelligence, single-mindedness and determination, and considerable eloquence. And he knows his military history. I was incautious enough to make some comment about the upheavals in Russia, and the recent Russo-Japanese conflict, compared to the relative calm of the European empires. Calm? he said. *Calm?* The calm of Alma, Balaklava and Inkerman? The calm of

Balichiao in 1859, where Lord Elgin's troops annihilated ten thousand Chinese, including the entire Mongolian cavalry, so that all of China would be open to British trade? That is the real British specialism, he said – war for export. At Isandhlana in 1879, 150,000 Zulu overwhelmed 1,600 British troops. At Ulundi, in the same year, your Lord Chelmsford had his revenge, and with 25,000 defeated 23,000 Zulu under Cetewayo, and burned the Zulu capital to the ground. I am impressed by this command of military detail, I said. There is more, he said. In 1899, your Colonel Wingate, with only 3,700 British and Egyptian troops, beat 45,000 Dervishes at Om Dubreikat, but of course he had the indispensable aid of the Maxim gun. Some thousands of miles to the south of there, and almost on the same day, Lord Methuen defeated 8,000 Boers on the Modder River, but a fortnight later 3,500 Boers beat 7,500 British at Magersfontein. Then you did it to them again at Colenso four days later, three weeks after that again at Ladysmith, and then they surrendered after the terrible poundings you gave them at Paardeberg, Pieter's Hill, the Zand River and Pretoria. Ah, the peace-loving British! Will you have more tea? he said. I will, I said. I forgot, he said – 1897, your punitive expedition against Benin. Admiral Sir Harry Rawson captured, burned and looted Benin City, with the result that the priceless bronzes of that people were largely destroyed or dispersed. Homes, palaces and religious buildings were torched. This is British peace, British calm. Please do not, as a loyal Britisher, pretend a horror of violence, Mr. Holmes. You have silenced me on that subject, I said.

After tea we took a walk along a pleasant little river in Islington, and I took this not-quite level photograph of Ulyanov. He gave me a charming letter of greeting and congratulation from a fellow-revolutionary of his, Miss Alexandra Kollontai, now in exile in Germany.

She was previously a librarian in St. Petersburg, and obtained Watson's stories for her library.

Thursday, 25th May Hawthorn is now bursting into bloom everywhere. Cold north and east winds, bright sunshine in the middle of the day, but cold, sometimes even frosty, nights.

Sowed cinnerarias.

Sunday, 28th May In church this morning for the christening of the first grandchild of our neighbour Lieut.-Colonel J. W. A. Watts-Silvester. During the service I saw a striking sight – in the front pew, an old gentleman from Hankham in a white embroidered smock, worn with corduroy knee britches and gaiters.

Thursday, 1st June The drought continues. A month has passed in which we have had barely enough rain to lay the dust. The farmers' prospects of a good hay crop are growing less as the weeks pass by, and a light hay crop, I am told, betokens a short honey crop. The annals of leafy June will – if a bountiful rain comes – soon raise a paean of praise among us bee-keepers.

In early Summer I take care to have drinking troughs near the hives. These are the old blue and white soup plates brought from Baker Street when we moved, and they stir memories of eating with Watson. The plates are filled with small stones, and when kept well supplied with clean water they keep the bees from visiting ditches and ponds where the water is less clean. This assists in maintaining their health when they need so much water in Spring.

There was excellent comedy in the village this afternoon. A swarm of bees got into the pillar box by the green.

Thursday, 15th June The bee-season still delays its coming. More than a third of June has passed without a good bee-day in the whole twelve. The fields are covered with blossom but the weather is so unpropitious that the poor bees cannot leave their hives. Rain falls on most days, and when it is not raining there is a leaden sky overhead with cold NE winds prevailing.

Much hoeing of beds and borders. Roses watered with manure water.

Wednesday, 21st June The solstice, and astonishingly light at 9.30p.m. Earlier this evening, I was stooping to pack away some bee tools in the garden when I heard a deep buzzing in the air. I looked up and saw a swarm – not from my hives – flying above me, circling lower and lower. I ran to get a brood box with frames, and in that time the swarm began to settle on the low branch of a pear tree outside our kitchen window. Against the light sky I could see the scout bees, highly excited, dancing at the cluster, flying up to the branches and then down again, encouraging the swarm to come down. Within ten minutes the entire swarm was clustered in my brood box, with their virgin queen.

Wednesday, 28th June A letter came this morning from the Grand Hotel, Eastbourne. M. Claude Debussy is there, and with him Mme. Emma Bardac, who is pregnant. I know him from some years back when he was leading a tempestuous life in the rue de Londres. I shall call on him.

Thursday, 29th June Beautiful weather during the past fortnight has redeemed the bee-year!

Wednesday, 5th July Weather conditions here are most favourable, and in strong contrast to those of last year. The limes are just coming into full bloom, and white clover is flowering abundantly. Thanks to the copious warm rains lately, there is every prospect of a grand yield from these sources.

Fiona has divided the garden pinks, and sowed perennial campanulas.

Thursday, 13th July Yesterday I went to see if there were any signs of swarming, as the day was exceedingly hot; but I found the bees in all hives busy at work, and not a sign of swarming. I was just coming away, when all at once I heard a tremendous buzzing, and looking round I saw the bees rushing out of one hive in their thousands. Not wanting this stock to swarm, I picked up a small pan of water that was handy, and threw the water into the mouth of the hive. That dousing stemmed the outrush, so I got more water and swished it up among the bees circulating in the air above. I did this like a veritable Whirling Dervish till most of them returned to the hive; I then took the lower drawer out, to give the bees more room. They hung onto the bottom of the combs just like a swarm on a branch. I then opened the hive and cut out all the queen-cells (ten

47

in number), and put two new frames of foundation in, and gave another rack of sections. I then used the smoker pretty freely and drove the bees up among the combs. Jas. Hiams stopped by just as I had finished, and he told me 'they would be out again soon'; but I am pleased to say his prophecy did not come true.

While no shade is necessary in the Spring or Autumn, shade during the heat of the day in mid-Summer is very desirable. In my own case, the shade of my cherry and apple trees seems satisfactory, and my five hives stand four feet from each other under the boughs, facing SE. These are as many hives as I care to look after. In June last year I had all four hives promising well but about the beginning of July I found one of them had lost its queen, and had to buy another. This hive will be strong before going into winter retirement, but will give no super honey this season. The other hives have given me about 80lb. Under each hive are bricks which act as a stand, with a piece of board leaned against the front of the hive bottom for the bees to run up on. Bees coming in heavily laden during the honey flow often drop to the ground some distance from the hive and are unable to rise into the air again. It is thus of considerable advantage to have a way they can crawl into the hive. With this in mind it is well to keep down grass and weeds in the neighbourhood of the apiary. I cut mine myself. A neighbour, William Tompsett, uses his prize sheep for this purpose, and also his ducks.

Thursday, 20th July Tomorrow is the last day of the school year, and this year it falls to me to give the pupils their prizes, and make a short speech. This dubious honour, I am told, is conferred on older inhabitants of the village more at less at random by Mr. Drew, the schoolmaster. The prize-winners will have performed well in school work, or recited a number of the Collects from memory, or simply gained a large number of attendance stamps during the school year. I intend to hold their attention by telling them the story of *The Hound of the Baskervilles,* though what moral I should draw from the tale I cannot, at this moment, say.

Friday, 28th July I have had two swarms this year; one of which unluckily flew away yesterday. This swarm issued at 7a.m. They hung on a bough till half-past one, when I came out to run my eye over the hives and saw one that looked quiet about the entrance, which made it clear a swarm had gone. I was just in time to see the swarm leaving the bough. After flying around they made off. I followed across three fields till, on coming to a belt of trees, I lost them.

Finished *Gargantua and Pantagruel.* It is extravagant, preposterous, occasionally shocking, always amusing. And by far the longest book I have ever read.

Friday, 5th August Yesterday with the Mintys in the automobile to Fishbourne, where Minty's sister lives. We had lunch and a walk afterwards through the fields. Everyone believes there is a grand Roman villa buried thereabouts, and a farmer called Fuller gave us several pieces of coloured mosaic tile that he has ploughed up.

Monday, 15th August I met M. Debussy yesterday in his suite. Why he has chosen to stay in Eastbourne I cannot understand. It is a jumble of bleached facades, plate glass, gold lettering, awnings, posters, flowerbeds, brass bands, crowds and traffic. He was engaged in correcting the proofs of a work called *La Mer,* which is three symphonic sketches. He told me he intends to arrange them for piano with four hands. I looked at the score, and told him that although it was highly original, and painted from a rich and dramatic palette, its extroverted emotionalism was not to my own taste. He gave me invaluable advice on my string arrangement of the *Goldberg Variations.* His divorce from Texier came through on 2nd August.

Tuesday, 23rd August Layered carnations now the flowers are over, sowed mignonette and, advised by Bill Frusher, removed seed pods from sweet peas to ensure continued flowering.

Thursday, 21st September That good man, Dr. Thomas Barnardo, died this Tuesday gone, at the age of sixty. His heart gave way. From how many criminals has London been spared by his life's work? Many more than I can claim. I heard that 100,000 children have been rescued from the streets since he began his work. I never met him, but I shall attend his funeral.

Planted narcissi, hyacinths, tulips and crocuses before squalls of rain drove us inside.

A messenger boy has just brought a telegram from someone called Frank Kornblum, requesting a consultation. I have sent a reply saying I shall be at home next Thursday morning.

Sunday, 24th September It appears that the weather has slipped and we seem, so far, to have missed out on the promised Indian Summer. The bees have not gone into a cluster but they have slowed and are not taking food down. It looks as if they are about to get settled for the Winter. If that is right, then I must hurry to provide the recommended stores to carry them through.

This morning the church was very prettily decked out for Harvest Festival with hay rakes and crooks, sheaves of corn, fruit, vegetables, and Michaelmas daisies. I love the hymn sung on these occasions – All is safely gathered in, Ere the winter's storms begin.

Wednesday, 27th September This morning I joined many thousands of people from the East End in the funeral procession of Dr. Barnardo as it went along Commercial Road, the Stepney Boys' Band playing solemn music at the head of the cortege. Blinds were drawn all along the route, and business suspended. We walked between dense crowds of hushed and reverent onlookers, the men, even the roughest, with bared heads, and children weeping. We stopped at Liverpool Street, and the coffin, with a small party, went on by train to Barkingside for the interment. Fiona and I have sent a letter of condolence to Syrie.

William Tompsett, I have found out, is an expert mushroomer, and, this being nearly the right time of year, he has offered to take us in the next few weeks to Abbot's Wood, near Hailsham, to see what we can find.

Friday, 6th October The Kornblums have been, and I intend to record what happened in detail. Yesterday morning, while I was in the front garden enjoying the mild sunshine, an astounding automobile drew up at our gate. It was a vision of gleaming black metal, brass lamps, vast yellow wheels, maroon padded seats front and rear and, most astounding of all, a chauffeur clad entirely – I mean, from neck to foot – in red leather, and wearing a white cap. He was a young Negro, who sprang out of the machine to hand down first a woman, then a man, whom I at once identified as Americans.

'Mr. Holmes!' called the man, pushing open my gate and striding towards me, hand outstretched. 'Frank Kornblum, of Newport, Rhode Island! Proud to make your acquaintance! May I present my wife, Belle, and say what a lovely, lovely home you have here!'

I had Pearl take the chauffeur to the kitchen while I led my visitors round the house to the orchard, where Fiona was sitting. Mrs. Trench brought out coffee and pastries, but the visitors would have nothing but lemonade. I showed them my hives, from a safe distance, Fiona spoke about her mother and sister in Chicago, and we heard of Mr. Kornblum's companies that make steel rails for the American railway system, and how he survived, thanks to God's special providence, the depression of the 1890s. I asked about the car.

51

'A Buick Tourer,' said Mr. Kornblum. 'Model G. Just a year old. Two horizontal cylinders, double opposed, water-cooled. Single chain drive, two forward speeds, one reverse. Jump spark ignition with dry cells, planetary transmission. It came over with us, and it'll go back with us. I'll change it for a four-cylinder next year. Now, Sherlock – if I may call you Sherlock? – and Fiona! – you'll be wondering why we're here, disturbing your tranquillity! Belle, step forward, please.'

'Well,' said the lady, 'this is our situation. We have a daughter, Agnes, who's twenty one. She's here in England at the moment, just returned from Italy, where she's been for two months. We wanted her to see the wonders of the ancient world.'

'In which part of Italy has she been?' I asked.

'Near Naples,' said Mrs. Kornblum, 'in a wonderful palace, she says, with gardens, halls and terraces you can hardly imagine. It belongs to a young man called Fabrizio Colonna di Paliano, a Prince of Summonte, whatever that means, who's very kind. There seems to have been one long house party there this year, with lots of young people coming and going, which is all very well, of course – in fact it's just what we hoped for her, and she's had a great time, except...'

She paused, and sighed.

'...except...'

'Except – she's got engaged to be married,' said Mr. Kornblum. 'She's got a ring with a diamond the size of an ice cube.'

'Is she engaged to the Prince?' I asked.

'No, not him. It's an English nobleman, a Viscount, very young, who proposed to her, who was also there on holiday. He's twenty two, his name is George Ridley and he lives near here – or says he does – at Ridley Hall. And in Portland Place, in London. And in Scotland.'

'Why did you say "or says he does"?' I asked.

'Because – ' he began.

There was a loud honking noise from the other side of the house, and I went to see what it was. A group of my fellow villagers, including many children, had gathered round the car and were gazing on it as if it had descended from a distant planet. One of them had pressed the bulb of the car horn. This had caused the chauffeur to come out into the front garden, and his sudden, splendid appearance on one side, and mine on the other, caused a sensation, with many of the onlookers backing away to the other side of the road. I went back to the orchard.

'There's a book you can get in America, Sherlock,' said Mrs. Kornblum, 'called *Titled Americans*. It came out – what? – ten, fifteen years ago? It's not a big book, but it's a mighty interesting one. It gives a list of young American women – all rich girls, like our daughter – who've married "foreigners of rank". That's how they put it. But it also gives you a warning about phoney foreigners who don't have any rank at all, but pretend they have, and they don't have any money, but they pretend they do. It says there are hundreds, maybe thousands, of Princes and Princesses in Russia, all with fancy titles, who are driving cabs, working as waiters and waitresses, and – why, there's even one, it says, who wears tights and spangles and rides bareback in a low-class circus! A Princess! And in Italy, it's just as bad. Did you know the Pope will make you a Count for $5,000 a time, provided he's paid in gold? And a Duke for $16,000? They have Dukes like Lincoln State Park has mushrooms.'

'I assure you that in Britain we are more thrifty,' I said. 'I do not believe there are thirty Dukes in this country and Ireland combined. Our House of Lords has fewer than six hundred peers all told, and many of those hold titles which become extinct on their death. But, even so, I can tell that you are concerned about young Ridley's *bona fides.* You say they are engaged. Have you met his parents?'

'We're to meet them next week,' said Mr. Kornblum. 'But we're scared of being bamboozled, Sherlock. I'm not ashamed to say it. How do we know if the parents - if they *are* his parents - are what they say they are? In America there's not a man can fool me in business, but over here nothing's clear, everything's not what it seems, and we're like the babes in the wood.'

There was another loud *parp-parp* from the front of the house. Fiona went to see what was happening.

'And how do you think I can help you?' I said.

'Well, aren't you the greatest investigating detective in the world?' he said. 'Belle and I want you to investigate him for us, and I pay well, I promise you. Frank Kornblum hires the best, and pays the best, believe me. I'm a millionaire a few times over, Sherlock. I know in England you don't talk that way, but where I come from, we do. I want you to snoop for me, and snoop good. I'll understand if you tell me to get out of your garden and back to Rhode Island, but, on the other hand, I figured you might just say yes, if I put it to you straight.'

'But I would have only a week, or less, to do this... snooping.'

'That's right, and I'm sorry about that, but that's the way it is.'

Fiona returned. The chauffeur, she said, was driving Mr. Kornblum's car slowly up and down the lane with at least twenty children clinging to it, having the time of their lives. I asked the Kornblums to excuse us while we walked down to the hives, and I told her what was proposed. She encouraged me to take the commission, reminding me that the Soup Kitchen needed new guttering and drains, and decent copper pots and pans. And how good it would be if the Kitchen had no fuel or lighting worries for the foreseeable future! We asked the bees what they thought, and took their loud, contented murmuring for assent.

'I shall do my best for you, Mr. Kornblum,' I said when we rejoined the Americans. 'If I can give you hard evidence one way or the other, I shall require a fee of $1,500. If I cannot, there will be no fee.'

'And expenses?' he said.

'I shall not charge for expenses in this case,' I said.

'Hey,' he said, rising and extending his hand, 'you have a deal, Sherlock!'

Saturday, 7th October I set to work as soon as they were gone with a last honk of the horn, a salute from the chauffeur, and ragged cheering from the children by my gate. I keep an up-to-date Debrett's, and a *Who's Who,* in the house. Viscount Ridley was born in 1884, on the 31st October, to be precise, and is an only child. He was educated at Eton College and at Merton College, Oxford, where he apparently showed promise as a zoologist. The republican in me was pleased to see, in passing, that Merton was the only

Oxford college to side with Parliament in the Civil War. He is intended shortly for the Royal Military College, Sandhurst. His father, Thomas, is Lord Colney, formerly a Captain of the Rifle Brigade. George's mother is Marchioness of Glentworth, formerly a Lady-in-Waiting to Queen Victoria, and a second cousin of H. R. H. Alexandrina, daughter of H. S. H. Joseph, the late Duke of Saxe-Altenburg. She is an heiress in her own right, and her jewellery alone is estimated to be worth £100,000. The entailed estates amount to 7,000 acres in Sussex and Wester Ross, yielding an income of £65,000. So far, so good.

Monday, 9th October Returned this evening from London, where I lunched, at a place I cannot name, a man whom I cannot name, who is a clerk at a bank I cannot name. This thrice-nameless informant – whom I met a little over ten years ago in connection with the dreadful case of Crosby, the banker – was able to vouch for young George's family's solvency, and for the family's freedom from financial taint or suspicion.

Wednesday, 11th October Fiona and I are returned from a most pleasant day in Oxford, where £5 paid to one of the porters of Merton College bought me the intelligence that George the undergraduate was as pleasant as one could wish, tipped well, and indulged in exactly the liberties and misdemeanours expected of a wealthy young aristocrat, and no more. This investigation is the most tranquil I have ever conducted.

Sunday, 15th October Well, we are back from a two-day cycling expedition – on a German tandem, no less, hired in Bolney, and in excellently mild weather for the time of year. We have explored the Burgess Hill area, where Ridley Hall is located, and come back with a most astonishing piece of information. In the village we first tried the Post Office where we learned, by asking if the young man had recovered from his illness, that he had never had a day's illness in his life, except of course the usual things as a little child. We then tried the village public house, The Royal Oak, one lunchtime. On my freely dispensing pale ale, stout and cider, we had good reports of Lord Colney's relationships with his tenants, of the happiness of the family unit, and the complete absence of scandal from its recent history.

Then we came to the big house itself, which we infiltrated in the following manner. We cycled to the gatehouse, where Fiona became faint in a most convincing fashion. I asked for help in the Lodge, and she was helped to the house, with me following.

While she was being put in a chair in the Servants' Hall and given water and *sal volatile*, with the promise of a good cup of tea to follow, I executed my part of the plan. I had brought with me, in a pannier on the tandem, a basket containing workman's clothes and a bag of plumber's tools. I now donned the clothes behind an outhouse and slipped upstairs, prepared to say, if challenged, that I had come to fix a leaking tap, but could not find anyone to tell me where it was. I had no clear plan except to gain some feel for the place; my report would be incomplete without a reconnaissance of where its subjects lived. I was slightly ashamed to be sneaking about someone's house, but I fortified myself by thinking of new gutters, drains and equipment for the Soup Kitchen. The house was Queen Anne, with a fine central sitting room comfortably adapted to modern living. From its open window I looked down on a vista of lawns and cedars, and a party of young folk playing tennis. I wondered which of them was the young English heir, and which the young American heiress.

I reached the top of the house by way of the back stairs, and found myself on a landing in the attic with a door opening off it. By the door I noticed a hook in the wall, with a key hanging on it. Then there were footsteps on the stairs. I quickly stood in the space between a cupboard and the end wall. I heard the person take the key off the hook and unlock the door, and then there were quiet voices – a woman's and a young man's, I fancied – then the door was shut and locked again, and the footsteps retreated back down the stairs until all was quiet. Someone shut up in a room in a remote part of the house? I asked myself what could be the import of this for my investigation, and decided to act.

I knocked at the door.

'Is that you again, Mary Ellen?' said a young man's voice from the other side. 'Why don't you let yourself in?'

'I'm a plumber, sir,' I said, in my best Sussex accent. 'Come to fix a leaky tap. Is it in this room?'

'There's no tap in here,' was the reply. 'My water is brought up.'

'I'd like to check your pipes, sir,' I said. 'And the ceiling, for any marks of water coming through from the roof.'

'You can't come in here. No one comes in here except Mary Ellen, and sometimes my mother.'

'I won't come in, then,' I said. 'Will you allow me to unlock the door and just look into the room, from the doorway? To be on the safe side?'

There was a pause, and then, 'Very well, if you must. But be quick.'

I took the key, turned it in the lock and opened the door. Directly in front of me stood the young man. He was pale and drawn, with dark, frightened eyes. He wore a simple flannel nightshirt, but what made me start was a vast padded turban on his head.

'Well?' he said. 'Are there any marks?'

I glanced at the ceiling, my eyes quickly taking in the sparsely furnished room. The chairs were padded, there were mattresses instead of carpet on the floor, and the fire grate was protected by cushions.

'No, no marks. All's well,' I said. 'May I ask, sir, why you wear that thing on your head?'

'My fits,' he said. 'My fits. You know what fits are?'

'I do. I'm sorry if you suffer them.'

'Six, eight, ten times a day,' he said. 'The epilepsy. It's not a madness, you know. Julius Caesar was epileptic. It's not a madness. We fall down.'

'Are you a... member of the family, sir?' I asked.

'I am Viscount Ridley,' he said. 'Edwin, my father's eldest son. But because of my fits I live... secluded from the world. It is for the best. My brother plays the role that might have been mine. He does it very well.'

'He is your younger brother, you say, sir?' I said.

'He is. We are twins, but I was the first born. When we are older, he will look after me. Mary Ellen has looked after me from a baby, she and my parents look after me now, and one day George will look after me. You are sure there are no signs of water coming in?'

'There are none, sir,' I said. 'You are quite safe. I will leave you now.'

'You are a good plumber, I am sure,' he said. 'I will be pleased if you come back to see me at some time in the future. Please be good enough to lock the door and hang up the key. Goodbye.'

Reunited with Fiona, and on our way again, I learned from her that George, the supposed Viscount, was much loved in the Servants' Hall, that he was going into the Army because of family tradition, but was to be groomed after that to take over and manage the investments and the estate, the hunting lodge in the Highlands, and the house in London. His engagement was of course the talk of the Hall, and the imminent arrival of the fiancée's parents eagerly awaited. I shall not tell Fiona about my conversation with Edwin, at least for a while. She laughingly regretted that we would not be there to see the Buick, its spectacular chauffeur, and the Kornblums in their best clothes moving majestically up to the front door.

Monday, 16th October I shall keep the Ridley family secret from the Kornblums as well as from Fiona. There are times when the appearance of truth is more satisfactory, for all concerned, than the truth itself. I made my report this morning, received my Banker's Draft, and we wished the Americans well as they set off for Ridley Hall.

Monday, 23rd October The Hiams's have a medlar tree of great antiquity in their garden. They brought us in a bowl of the fruits this morning, which we had after lunch with cheeses.

Some unnecessary rearrangement of the shrubbery this morning at the whim of the mistress of the house.

Friday, 27th October On Wednesday we were in London for the first performance in England of the Fourth Symphony of the Austrian, Herr Gustav Mahler. We went as the guests of Freeman Freeman-Thomas, the MP for Hastings in the Liberal interest, and his wife, Marie. Henry Wood conducted. Wood's wife, Olga, sang a peculiar, ravishing song at the end. She was excellent – clean, clear, not operatic, child-like and attentive to the phrasing. Beautiful strings (including an excellent solo violin in the Second Movement) and woodwinds, biting trumpets, excellent horns, a shining harp. In the strings I noted the *col legno* and the *sul ponticello* very well done. We stayed at the Langham, and argued about South Africa till very late.

Monday, 30th October Planted cabbage, savoys and and coleworts in the kitchen garden.

An excellent mushroom expedition yesterday afternoon, in brilliant sunshine. We got plenty of ceps, chanterelles, bay bolete, shaggy ink cap and giant puffball, all edible and all delicious. I, of course, was also interested in the poisonous varieties, and brought home a beechwood sickener, a pretty pink and white confection that gives spectacular gastric symptoms; an orange scale head, which contains psilocybin and gives hallucinations as well as sickness; and, under some pine trees and looking like a crumpled brown paper bag, a magnificent, and utterly deadly, turban fungus.

Wednesday, 8th November I am fairly sure that my bees are now ready for Winter. The hives are arranged as I want them, all secure against the coming weather. It is a wonderful feeling to have so many new bees that should safely carry the entire colony through the coming Winter.

Tuesday, 28th November Walked to Friston this afternoon where I took this photograph of the fine old mill.

The cold weather is with us already. The frosts are quite serious, so the bees will now be in clusters and the queen will cut down on egg laying.

The revision and ordering of my criminal scrapbooks goes on. I find that sometimes I almost wish for bad weather, to give me the excuse to stay indoors and do this work. Few know the name of Mary Ann Cotton, but it may be that she is England's most industrious mass-murderer, with twenty persons close to her dying mysteriously, including three of her four husbands, eleven of her thirteen children, and various lodgers, lovers and neighbours. A further fifteen persons may also have died by her hand. She was tried for the murder of her second step-son, by arsenic poisoning, when the local doctor became suspicious and refused to sign a death certificate for the boy, and hanged in March 1873.

Sunday, 3rd December To London yesterday to see Watson and Rekha, and with them to a matinee at the Royal Court Theatre to see *Major Barbara*, a new play by Mr. George Bernard Shaw, who also directed. The actor who played Andrew Undershaft kept forgetting his lines. We had tea with Shaw afterwards. If he can be kept off the topics of vegetarianism, socialism and spelling, he is the wittiest man in the world. He made us laugh by telling us that, as part of his research into the Salvation Army for the play, he lustily sang *When the Roll is Called up Yonder* at a Salvation Army gathering.

Sunday, 10th December Jas. Hiams told me last night that when he was a boy Sussex farmers used to put sprigs of holly on their hives at Christmas. On the first Christmas, they believed, the bees in their hives hummed to welcome the Christ child. I told him that Ethiopians have a saying that Christ "is born from the voice of his father, as the bee is born from the queen", and they believe that bees once defended the throne of God. He was astonished to learn that Ethiopians are Christians, but pleased to know they hold bees in such esteem.

Wednesday, 13th December In freezing cold, applied coconut fibre on beds where the bulbs are, and around the most tender of the shrubs and roses.

Tuesday, 19th December Yesterday afternoon a boy came to tell me that a colony of feral bees had been discovered clustered in a hedge two and a half miles from here. At first I did not believe him, but he prevailed on me by dancing up and down in his agitation, and I went with him to ascertain their condition. The swarm must have been a very fine one indeed last Summer to allow of the great strength of numbers of the bees six months later. They had clustered on the north side of the hedge and built half a dozen combs of average size, with one end attached to a log. They were fully exposed to the north east wind, and there was no food to be seen. On my making a start to gather them, with the kind help of a local man, Daniel Icingbell, we had a warm reception, and had to retreat. The bees had, we found, enough food at the rear end of their combs to last them three or four more weeks, and what surprised us even more than their numerical strength, was that they had three combs well filled with brood and eggs. It was with the regret that only a bee-man can feel that I was unsuccessful in securing these bees and providing them with a comfortable home. Having driven us off, most of them, along with the queen, passed through a small hole in the log to a place where I found it impossible to reach them, and in consequence I had to leave them to the fate that, in any case, would have overtaken them later on. But in addition to my disappointment in having to leave the poor bees to perish, there was the great longing I felt to secure the queen, whose extraordinary vigour and prolificness was evident, and who would have been a great prize if only I could have caught her.

On the same day, hard though it is to believe, the bees from my three hives were busy taking in huge loads of pollen from the wall-flowers and ivy still in bloom in my garden; often there was such a merry hum in the neighbourhood of the hives that I could hardly believe it only wanted a week until Christmas Day.

Wednesday, 27th December We were at Spade House in Sandgate for Christmas, the guests of the Wells's, Herbert and Jane, with their two little boys, George and Frank. Wells gave us a beautifully inscribed copy of *Kipps: The Story of a Simple Soul.* It has already sold 10,000 copies, and been praised by Mr. Henry James. Great things are to be expected of it. I gave Fiona a pair of bright red calfskin gloves, and she gave me the ninth series of M. Jean-Henri Fabre's *Souvenirs Entomologiques,* which I had, I admit, asked for. He is a passionate advocate of the education of women, and as a result has been ejected from his professorship, and from his home, indeed, in Avignon. I have written offering my sympathy. Darwin called him "the inimitable observer", and I, too, bow the knee to his unique powers of observation. And to his good humour and gaiety, rare qualities in science.

Friday, 29th December For the last two weeks the apiary has looked dead in the bleak, dark weather. I am confident, though, that the hives are as secure as I can make them and well founded, too; roofs are weather tight; the hives stand firm on secure stands so there will be no rocking about in a high wind. I went along and hefted the hives. They feel as heavy as they were nearly two months ago; mouse guards are in place and all looks well. The bees should be snug and warm for the Winter. Nothing further can be done to help them; all is now in the lap of the gods.

Sunday, 31st December As I did last year, I have put together some notes which I hope may be useful to amateur bee-keepers. These notes are on my equipment.

The beginner will need a good veil and gloves, a suitable hive tool, a smoker and a bee brush.

Veil The globe veil, listed in nearly every catalogue, is a nuisance. Sewing silk tulle, brussels-net or mosquito netting to the rim of a straw hat offers protection but the netting catches on every twig and is easily torn. I use the Alexander veil, which is made from a strip of screen wire rolled into a cylinder, with a cloth sewn over the top and an apron at the bottom, which can easily be tucked under the coat.

Gloves Cotton flannel gloves with long gauntlets are for me the most satisfactory, and to be preferred to rubber or other heavy materials.

(A correspondent of mine, Mr. J. Martin, of Cuba, has recommended me an entire bee-suit, as follows: "In a clothing store I found what is called an engineer's suit – overalls and short coat, or blouse, made of blue and white checked cotton cloth, the whole weighing only 1¼ lbs. There are plenty of pockets fore and aft for pencils, jack knives, screwdrivers, queen cages and what-have-you.")

Hive Tool I have used several of the advertised hive tools, all perfectly satisfactory. The tool should have a sharp surface to scrape off burr combs, propolis etc., be strong enough to serve as a pry in loosening frames, and be about seven inches long, that is, small enough to handle quickly and easily. It should also have a hook at one end to help when lifting the frames.

Smoker The smoker is used to calm the bees down when one is inspecting the hive. The effect of the smoke is to make the bees think they may have to evacuate the hive, and so they gorge themselves on honey. This stupefies them and they become more docile. There are two excellent kinds of smoker on the market, the "Bingham" and the "Crane", and many indifferent ones. The "Bingham" is reliable and strongly made, but has a comparatively weak blast. The "Crane" gives an excellent blast; its only defect is that the check-valve sometimes becomes clogged with creosote, but this is only after the smoker has been used continuously for a considerable period of time.

As to size, I recommend the larger in both cases. Rotten wood is a good smoke fuel, although excelsior, cotton rags, hessian, pine needles, greasy waste or even rolled paper will do. Mrs. Richardson, of this village, dries rosemary and other aromatic plants, and says they produce the coolest and most calming smoke. Care should be taken not to use too much smoke. If one has gentle bees, a slight puff at the entrance and then another over the frames when the cover is removed will be sufficient. If the bees are inclined to be cross, a little more may be necessary. One must bear in mind that the use of smoke is disturbing to the bees, and every disturbance during the honey flow will be accounted for in honey stored.

Bee Brush The bee brush is used when one takes the hive apart. Bees invariably crawl onto the sides of the hive boxes, and the brush is used to gently clear away any bees that might otherwise be crushed.

Mouse Guard One additional piece of equipment that is essential, but only required in the winter season, is the mouse guard. A mouse can squeeze through the entrance to a hive and set up a cosy home for the Winter, with free honey to boot. A metal strip, drilled across with bee-sized holes and placed across the entrance, will keep mice out, but allow the bees to pass through in both directions.

1906

Wednesday, 3rd January To Birling Gap and walked on the beach on New Year's Day in bright sunshine. The previous night we attended a Salvation Army watchnight service in Eastbourne and then a noisy party in Burwash. Crowds, two dance bands, fireworks, tin horns and mayhem. I am instructed to say that Mrs. S. Holmes wore black silk with tucked sleeves and yoke, a black and white feather boa, and a grey bonnet with purple roses and aigrette, and Mrs. R. Kipling heliotrope and white muslin over heliotrope silk, trimmed with white lace, and a black hat with pink flowers and rosettes of chiffon. Both ladies had their palms read for 1906 by one Madame Zingari, and I was introduced to billiards by Lieut. Colonel Andrew Bethell and Major Anthony Mundy of the Royal Sussex Regiment, come over from the barracks at Chichester. In the early hours, I had the pleasure of meeting and talking with Mr. Harry de Windt, the explorer and author of *From Paris to New York by Land,* the tale of his stupendous journey through Siberia and across the Bering Straits. He told me the story of the purchase of Alaska from Russia by the United States, for the price of $7,000,000. Since that acquisition in 1867 Alaska has returned, he says, $100,000,000 in fisheries and furs, and $40,000,000 in gold and timber.

A kindly Mr. G. M. Doolittle, of Onandonga, N.Y., has read Watson's stories and sent me what he calls a "bullet-proof" vest, made of thirty layers of silk. He does not know that I have retired, and the parcel has been forwarded from Baker Street. He has included a pamphlet entitled *The Impenetrability of Silk to Bullets* by George E. Goodfellow of Tombstone, Arizona. I think the vest might possibly stop the relatively slow bullet of a black powder handgun, but not one from a modern weapon.

There was talk in the Tiger last night of a poltergeist making mischief in Hankham, a village some five miles from here. Apparently, a wealthy widow called Mrs. Morton recently moved into an old house there, and has found her vases and ornaments flying about. I was the only one in the Snug who refused to countenance the existence of such spirits, and was laughed at for my narrow-minded rationalism.

Sunday, 7th January A quiet celebration for our second wedding anniversary and my birthday yesterday. The Wellses came over from Sandgate for dinner and gave us an unusual print by an artist called William Blake, not previously known to me, showing a scene from Dante's *Inferno*. I gave Fiona an enamel bee brooch from Garrard's, with diamond wings and ruby eyes. She gave me a French folding magnifying glass in brass, which when folded fits easily into my coat pocket.

Monday, 8th January The Soup Kitchen in Eastbourne was re-opened last Saturday at 10.30a.m. We disposed of eight hundred quarts of soup, and six bushels of bread. All sorts and conditions of men, women and children attended, many looking as if they had not had a square meal for weeks. The heterogeneous assortment of vessels was remarkable. Some were orthodox jugs; others were corned beef tins, tins which had at one time contained toffee, wash-hand jugs and stone vessels of all kinds. Mr. P. Higgins, the chef, told me that last year no fewer than 18,836 needy persons were relieved with soup and bread in the town. That means something like 360 persons each week, in this one town. I met a poor man who had walked from Weymouth to Brighton, looking for work and finding none.

Tuesday, 16th January Sowed cauliflower, cabbage, spinach and lettuce.

I have been consulted on a case. Mrs. Morton, the widow at Hankham who was talked of a month ago as having trouble with a poltergeist, came over to East Dean this morning to seek my help. She is still being annoyed from time to time by crashes in the night, and recently by an upset cooking pot when no-one was in the kitchen. Like an old boxer who hears the ringing of a bell and gets up with his guard raised before he knows what he is doing, my immediate instinct was to look into it, but on a very few moments' reflection, I declined. She is a dignified and intelligent woman; I told her to talk to the police, and get them to look to some disgruntled servant or unkind neighbour.

Saturday, 20th January On entering my garden on the morning of Thursday last I wondered whether it really was the third week in January, or a warm day in May. What with the warmth, the bright sunshine and the hum of bees around me, it was more like Summer than Winter. The bees were out in large numbers, revelling in the opportunity afforded them for an airing flight. A bee-keeping friend and neighbour, William Loveday, told me he had seen numerous bees buzzing past him out in the fields a mile at least from any apiary.

Friday, 9th February Up at six to see the eclipse of the moon, and saw the first contact with the shadow, but then clouds came over and hid the total phase.

Monday, 12th February I have been, after all, to Hankham, and that visit proved so extraordinary, so terrifying, and so nearly fatal, that I cannot convey the matter properly in a short journal entry. I wish Watson were with me, to do justice to the telling of such a story. I will write a full account of the case over the next week or two, and present it to my readers when it is finished. Till then, I will resume my usual humdrum journal entries. Suddenly the humdrum seems very acceptable, compared to the nightmare events of the last weekend.

Tuesday, 13th February The General Election is over. The Liberals have triumphed and the Conservatives have lost more than half their seats. Campbell-Bannerman is Prime Minister, Asquith Chancellor, young Churchill Under Secretary of State for the Colonies. Eastbourne's representative is Hubert Beaumont, 1st Baron Allendale, of Eton and Balliol – but a Liberal.

I had a conversation in the Soup Kitchen with a Mr. Charles Booth, creator of *Maps Descriptive of London Poverty.* He lives with working-class families and records his findings in maps and diaries. His interesting contention was that Soup Kitchens, and the imminent Old Age Pensions arrangements for certain persons over the age of seventy, and other similar provisions this government plans for the poorest, are the only means to prevent socialist revolution in this country, such is the misery of so many men and women. This in the country that leads all others in pomp, and pride, and wealth!

Thursday, 15th February Under Bill Frusher's guidance, took chrysanthemum cuttings for late flowering, and put old dahlia roots into frames, to produce cuttings.

Tuesday, 20th February Today's entry is probably my longest in this journal so far, and it is certainly the most dramatic, and tragic. It is the account of an investigation I undertook just over two weeks ago, and have now written up. In the two-day climax of the case a man died violently, and my own over-weening confidence in my powers very nearly cost me my life. This is not a "Watson story" – he has a talent for story-telling that I do not possess – but I have tried to reproduce some of his effects, particularly the heightening of suspense which is gained by suppressing certain key observations made in the course of an investigation, only to reveal them at the end.

I was at my hives on Friday the 9th when Fiona came down the garden to tell me that I had unexpected visitors. I laid aside my veil and gloves and went into the house.

They were two – a woman in her mid-30s, I judged, in the black crepe costume and jet necklace that conveys the saddest of messages, and a man of about the same age, whose manner was agitated. He rose as I entered.

'Mr. Holmes?' he said, seizing my hand. 'From the bottom of my heart, I thank you for seeing us! I am James Morton, come over from Hankham. This is my sister, Miss Emma Morton. You have met my mother. She came to see you a fortnight ago.'

Miss Morton inclined her head, but did not meet my eye.

'Your mother did indeed call on me,' I said. 'I am sorry I was not able to help her. Forgive me, but I observe that your sister is in mourning.

'Our father,' said Mr. Morton. 'A sudden death, un-looked for, some while ago.'

'Please accept my sympathies, sir,' I said. 'And my sympathies to you, Miss Morton.'

The lady did not acknowledge my condolence.

'Thank you,' he said. 'Firstly, I must apologise for our coming to you like this.'

'No apology is necessary,' I said. 'I am well used to visits without preliminaries. But they do suggest some urgency in the visitor's affairs. Please sit, and tell me what brings you here.'

'Something appalling, Mr. Holmes,' he said. 'She has lived there for nigh on six months, and she has told you herself of this poltergeist nuisance that has plagued the house since she moved in. But now, my God! – '

His voice broke.

'Please give me just the facts,' I said. 'As simply as you can.'

He composed himself.

'Last Tuesday,' he said, 'my mother was resting in her upstairs sitting room, after lunch, when her Staffordshire spill-holder fell from the mantelpiece and smashed on the slate below. She jumped up at the noise and examined the pieces. She was at an utter loss to understand how this had happened. The windows were closed, as they always are, and she was alone in the house. Her housekeeper, and only servant, Mrs. Stiven, had gone to the village shop. For the first time, this poltergeist, or whatever it is, has struck in the same room where she was. Then she heard a loud crash from the kitchen below. She went downstairs and found that six dinner plates had fallen from their shelf in her Welsh dresser and were lying in fragments across the floor.'

74

'They were a wedding present,' said Miss Morton, looking at me for the first time. 'With her and my father's initials entwined on every one.'

'Were either of you nearby?' I asked.

'My sister lives in London,' said the brother, 'and I in Nottingham. My mother sent me a telegram later that same day, asking me to come at once. I contacted my sister, travelled to London, and we came down together on Wednesday, from London Bridge.'

'Pray continue,' I said.

'My mother put her coat on and went along the lane to her nearest neighbour, Mrs. Thwaite,' he said. 'Mrs. Thwaite came back with her and helped her clear up the mess. She made her some tea to steady her nerves, and sat with her until Mrs. Stiven returned. Then she suggested to my mother that the two of them take a cab into Hailsham, to inform the police.'

'I suggested that course to her when we spoke,' I said. 'Your mother has no telephone?'

'No,' he replied. 'Mrs. Stiven went to the house of the village cabman, and he brought the cab round.'

'And then?'

'In the police station they were laughed at, and told it was not police business. They went to the Post Office and sent the telegram to me, and then they decided – you may smile at this, Mr. Holmes – to go to the Free Church, which my mother occasionally attends.'

'An understandable decision, given the circumstances,' I said.

'The minister there recognised her. Unfortunately, he offered her nothing by way of comfort except to say that he did not believe in poltergeists, and that he was a busy man. So the ladies went along to St. Saviour's, where a young curate said more or less the same. Finally, they went to St. Michael's, the Catholic church, where the priest said a prayer for our mother, made the sign of the cross over her, and bade them good evening. They sat together in the kitchen when they returned, and had some supper, and then Mrs. Thwaite suggested that our mother come and spend the night with her, as she was too frightened to stay by herself in a house that was haunted. They went upstairs together to our mother's bedroom to get her night things, only to discover, when they opened the door of the bedroom, that dresses and other clothes were out of her wardrobe and scattered across the room, some of them torn.'

'Torn?' I said. 'Are you sure of that?'

'Quite sure,' said Miss Morton. 'She showed them to me.'

'An unusual poltergeist,' I said, 'that not only breaks objects and scatters things about, but tears them, too. What did your mother do then?'

'She went straight to Mrs. Thwaite's house, and has remained there for the last three nights. She is a strong woman, Mr. Holmes, but her nerves are quite gone. Mrs. Stiven has gone home to Wilmington, and the house is locked up.'

'May I ask why you did not come to me yesterday?'

'My sister and I were both tired after our travels. I especially, as I had had a twelve-hour journey from Nottingham. We did not get to Hailsham until 7p.m on Wednesday. We spent yesterday resting, and trying to comfort my mother.'

'And where are you staying, Mr. Morton?'

'We are at Bellamy's Hotel, in Hailsham,' he replied.

'So, what do you imagine I can do for you?' I said. 'Apart from offering you refreshment, which – forgive me – I have signally failed to do.'

I rang the bell, and asked Pearl to bring us tea.

'You are our last resort, Mr. Holmes,' he said. 'I read in the newspapers a while back that you had quit London and come to this part of the world, but I thought no more about that until this – this latest trouble. I persuaded my sister to come with me today.'

The sister looked up at me with tears in her eyes.

'I told my mother,' she said, 'that as the police and the church can do nothing for us, perhaps Mr. Holmes will relent on our entreaty, and help us.'

'I am touched by your distress,' I said. 'I will come to your mother's house, but I must tell you that it is very unlikely that I shall be able to help you. I have no experience in matters of this kind. A malicious person might be able to get into your mother's kitchen to break the plates, but not into her sitting room while she was in there. How long has Mrs. Stiven been with your mother?'

'Since we were children,' he said. 'She came with her down here because she has nowhere of her own. Why do you ask?'

'She was in the house while your mother and Mrs. Thwaite were in the town,' I said. 'She could have scattered and torn the clothing.'

'That is simply not possible, Mr. Holmes,' he said. 'She adores our mother, and is the very soul of kindness and goodwill. If you could meet her, you would see at once that she could never do such a thing.'

There was nothing more to discuss, and I arranged with him that I would meet him at the house in Hankham the following afternoon. Now, if Watson were writing this, he would suppress what was going through my mind at this point, to keep his readers guessing, but I choose to disclose my thoughts here and now. I could glean nothing germane from the sister's appearance. As for the brother, his clothes were of good quality, and his demeanour that of a gentleman, but I had observed dark lines, that might have been earth, under his fingernails, which washing had not quite removed. There were also some tiny abrasions on the fingers. Had he been digging? His shoes – from Lobb, of London, by the look of them – had traces of earth beneath the instep. And there was something about his features – his forehead and his eyes, particularly – that intrigued me; I felt I had seen them somewhere before. Of two things I was quite sure, however: there are no poltergeists in east Sussex, or anywhere else, for that matter, and on Wednesdays there is no train from London that arrives at Hailsham at 7p.m. The last arrival is at 4.15p.m. When my visitors had left I spent some time consulting my one-inch-to-the-mile Ordnance Survey map for the Hankham area, packed certain items into a haversack, and wrote a note to Fiona, explaining my forthcoming absence for one night.

It was just after 5p.m. the following evening when my cab pulled up by a tall yew hedge, a good half a mile from the centre of the village of Hankham. A light rain was falling. I went through a creaking iron gate and stood looking up at the house. It was a gloomy building with narrow windows in the Gothic style, and mantled in a thick, dark ivy. Against the left wall stood a greenhouse full of overgrown plants whose leaves pressed at the glass panes. Almost at once I heard the sound of feet on gravel, and then Mr. Morton was coming round the side of the house and across the lawn to greet me.

'The house is neglected, Mr. Holmes,' he said. 'As I told you, my mother is almost a recluse. She pays no attention to the upkeep of the place. Please follow me.'

He led me round the greenhouse and along the side of the house. Over his shoulder I got a glimpse of an extensive garden, bounded by a low stone wall, with trees and fields beyond. We went in through a side door and into a corridor which led to the kitchen. There was the dresser, still with two rows of plates on view, but with one shelf bare. I examined the dresser, and the pieces of broken plates in a sack in the corner, without seeing anything of significance. Then we went upstairs and into a sitting room whose furniture and decoration were all in the taste of fifty years ago. He pulled back the curtains to allow me to inspect the over-mantel and fireplace, where again I drew a blank, but as we were about to leave the room I noticed a rectangle of lighter tint on one wall, which seemed to indicate that a picture had been removed from its place after hanging there for some time.

At that very moment we heard a colossal crash outside the room, the sound of splintering wood and breaking glass, and the curious striking of a bell. We both ran to the door. On opening it we saw that a majestic long-case clock had fallen from its position at the top of the flight of stairs leading to the second floor, and hurtled down onto the tiled landing where we stood, knocking out bannisters and tearing the stair carpet as it came.

I ran up the stairs. It was too dark to see anything on the landing there.

'The gas is turned off,' said Mr. Morton. 'I'll get you a candle from the kitchen.'

When I had the candle in my hand I examined the wall. No brackets had held the clock in place. It had simply decided, apparently, after fifty years or more without moving, to lean forward and plunge down headlong. All it had left, strangely, was the very faintest trace of a woman's perfume on the air.

'I must confess, I am trembling!' he said, as I came down. 'Now we have seen it with our own eyes, Mr. Holmes!'

'Not quite,' I replied. 'We heard it. I would like to look into those two rooms on the landing upstairs.'

'They are locked,' he replied. 'My mother will have the keys somewhere, though. I can ask her for them.'

I declined his offer; the point was to look into them immediately, not later. As we made our way out of the house and towards my waiting cab I told him, with apology, that there was nothing I could do for his mother. I was sorry for her, but I could not discern the cause of the mysterious phenomena. But, I said, I would return the following morning to see what else might fall, or break, while I was there.

A quarter of a mile down the road I told the cabbie to stop. I got down, with my haversack, and paid him off. It was my plan to circle back through the fields to a coppice that stood behind the Morton house, there to make myself as comfortable as I could till nightfall. I felt sure my vigil would be rewarded, somehow. It was a damp walk, and a damp hiding place I found, but – the game was afoot, once more! I hunkered down with an unexpectedly cheerful heart and allowed myself, as it grew dark, a nip of whisky and a comforting pipe. At midnight I rose, somewhat stiffly, made my way out from the trees and across a stretch of rough grass towards the rear wall of the garden. I got easily over the wall, and stood beneath a vast rhododendron to watch the windows at the back and on one side of the house.

I must have stood there for an hour, with my faith in my instincts beginning – I freely admit it – to weaken slightly. Then, that faith was vindicated, and I had my reward; someone was in the house. Dim light from a candle appeared in a first-floor back room, then disappeared, and then flashed from the kitchen window directly across from where I stood. Then all went dark again. *It can only be Morton,* I thought, *or perhaps the sister, preparing new breakages for me. But why?*

At the side door I took from the haversack my skeleton keys, exquisitely filed lever-lock keys made for me in '87 by Peter Marshall of Paddington Green, the cleverest locksmith that profession has seen in a hundred years. They would open any domestic door in London, he once told me; a side door in a country dwelling was child's play for them. In a moment I was inside the house. I began to feel my way down the corridor, past the kitchen, and then, with my eyes fully adjusted to the dark, I saw a flight of steps going down to a cellar, and there, beneath the cellar door – a line of light.

I went down noiselessly and put my ear to the door. Someone in the cellar was digging; the rhythmic sound of spade going into earth was unmistakable. I took the life preserver from my bag, wishing I had Watson by my side, and threw open the door.

A single candle on a table cast its light onto Mr. Morton, who was down in a waist-high trench in the floor, his back to me. He turned and saw me standing in the doorway.

'Mr. Holmes!' he exclaimed. 'What in God's name are you doing here?'

'The very question I had in mind to ask you,' I said, stepping towards him. 'This is strange work for this time of night.'

'Strange, indeed,' he said, 'yet, with a purpose.' Then his expression changed from one of astonishment to a broad smile. At that moment I heard the door click shut behind me. I turned and in the flickering light saw two women standing there – Miss Morton, gazing at me calmly, and next to her a woman I recognised at once as Mrs. Morton, wearing mourning like the younger woman, and holding in her right hand a short-barrelled Webley revolver.

'We knew you would come, Mr. Holmes,' she said. 'You could not resist. We are familiar with your methods.'

For a moment I felt my senses reeling. It was essential that I steady myself at once.

'Mrs. Morton,' I said, 'I did not expect to meet you here. You must excuse – '

'My name is not Morton,' she said. 'Mine is a name you know of old. I am Lucy, widow of James Moriarty. Yes, you may well give a start, Mr. Holmes! These are the children whose father you cruelly took from them. And see, my son has prepared a grave for you! You will shortly lie there. Then, though my dear husband's body was never found, his soul will be at peace at last.'

It came to me in an instant, why her son had looked familiar. He had Moriarty's deep eyes and domed forehead. How could I not have realised that before?

'Ah,' I said, 'then it was Professor Moriarty's picture that had been removed from the wall upstairs, in case I should see it. I congratulate you, Mrs. Moriarty. I have walked into a pretty trap – one whose setting is worthy of your husband. You came to this house, I take it, in this village, and created the story of the poltergeist, with the express purpose of luring me here.'

'I did,' she said. 'And I can almost believe it was the sheer intensity of our desire to be avenged that brought you to us tonight. My husband was a man, Mr. Holmes, worth twenty of you, yet you snuffed out the life of that great spirit without a regret.'

'I assure you,' I said, 'that I have many times thought of your late husband with respect for his powers, and with regret that it had to come to killing between us. When we met above those fearsome falls in Switzerland he showed a generous and courteous spirit towards what he imagined was a defeated foe, and kindly allowed me time to write what we both thought would be my last letter to my friend, Dr. Watson. But, in the world in which we moved, it was inevitable that there should one day be a final reckoning.'

'Kill him, mother,' said the daughter, quietly. 'I cannot bear to hear one word more from this fiend. Kill him!'

Her mother began to raise the revolver towards my heart.

'I should point out to you, madam,' I said, nodding towards the weapon, 'that nothing can be achieved with a weapon whose safety catch is on.'

83

Though the widow of a practiced murderer, she was not one herself. She could not control the instinct to look down, and that glance downwards gave me just the time I needed. I sprang to the table, extinguished the candle with my palm, and threw myself to one side. Three shots blazed out rapidly in the profound darkness, three appalling flashes of flame with deafening roars, and on the third a terrible cry came from the man who had been digging my grave.

'My son!' shrieked Mrs. Moriarty, and I heard the swish of her skirts as she hurled herself forward in the darkness towards him. The way to the door was clear. As I opened it, Miss Moriarty tried to clutch at me, but I was able to get up the stairs and gain the corridor, and then the safety of the garden.

I walked across country to Hailsham, and was at the police station before dawn. When we arrived at the house in the early morning light we found in the cellar the body of the son, shot through the heart and lying in the grave he had prepared for me. Of Mrs. Moriarty and her daughter, Emma, there was no sign.

So – what more is there to say? Even in the depths of the English countryside Moriarty's long shadow fell across me, but for a second time the gods who order these things decreed that I should survive. I am quite sure the cold flame of hatred still burns in his daughter's heart, and it may yet be that she stalks me again. It is not likely that she will read this, but if she does, I want her to know that I still regret my part in the violence that orphaned her, and I wish I could undo it; and I believe I can contemplate death at her hands, if it should come to that, with equanimity. Now to resume – I hope! – my tranquil village life.

Thursday, 1st March If typical wintry weather of the old-fashioned sort is to be taken as a fore-runner of good things in store for bees and bee-men, the outlook for the coming year is propitious. On Monday last we had a very severe snowstorm, with the result that our roads have been blocked all this week. As outside work is impossible I have had a busy time with saw, plane and hammer, accumulating bee furniture in readiness for the season's campaign.

February has come and gone, and from a bee-keeper's standpoint, has left little to complain of. I saw bees flying on a few occasions, but for the most part the month has been damp, dull and cold, with some half-dozen frosty nights. At such times quiet reigned in the apiary, which was no doubt the best possible condition for it; it had the beneficial consequence of saving both bee-life and stores. My stocks appear to be coming through well, and no signs of dysentery have appeared, which I was forewarned of by Jas. Hiams. He told me to leave the entrances open full width all Winter, and that seems to have had the right effect. Bees have been out in numbers during the last few days, and these cleansing flights are, apparently, among the best preventions of dysentery.

Saturday, 3rd March To the Frankensteins last night to dinner, where a Miss Muggeridge tried earnestly, over the steamed marmalade pudding, to persuade me to forsake the Italians and get Black British bees, our native bee. Apparently, they throw fewer swarms than other races of bee and have many important qualities evolved over thousands of years that make them entirely suited to our climate. They are prodigious collectors of pollen, she said, very hardy in Winter, not easily cowed by wasps and hornets, and excellent honey makers. But I like my good-natured Italians.

Tuesday, 6th March Today we received an ornate invitation to what we are quite sure will be the American society wedding of the year, to be held in the First Baptist Church in America, North Main Street, Providence, Rhode Island, U.S.A. Miss Agnes Kornblum and George, Viscount Ridley, will be united in matrimony by Pastor

Henry Melville King. We cannot go, but are sending two jars of my best clear honey. The present and future wealth of this young couple is almost beyond calculation, so the simplest of wedding presents is in order. We have requested a photograph of the happy event, to put up in the dining hall of the Soup Kitchen. I wonder what Edwin, the real Viscount Ridley, will know about it, in his attic hiding-place in Ridley Hall. I picture him unselfishly happy.

Thursday, 15th March To London on Friday 9th, to – of all places – Baker Street, for the opening of the Baker Street & Waterloo Railway. The line has been built by the Underground Electric Railways Company of London, and runs from Baker Street – still bustling unchanged, and seeming not to have noticed that I have left it – to Kennington Road. We boarded and left the trains through folding lattice gates at each end of the cars, these gates operated by gate-men who ride outside and announce the station names as the train arrives. Each station has its name picked out in tiles, in a particular colour. We had lunch at Grand Central Hall, Marylebone and listened to a speech by the Chairman of the London County Council, Sir Edwin Cornwall, M.P. By hansom cab the fare from Baker Street to Piccadilly Circus is 1s.6d., and the journey takes twenty minutes, traffic permitting; the same journey by underground railway is 2d., the journey takes seven minutes, and there will never be a problem with traffic. I fear for the cabmen that I used to rely on so much.

Yesterday to a reception at the National Gallery to celebrate the acquisition of *The Toilet of Venus* by Velàzquez. Fiona and I made our own small contribution to the National Art Collection Fund that raised the asking price of £45,000. This is by no means my favourite painting of his – in that regard I waver between *Las Meninas* and *The Surrender of Breda* – but I revere Velàzquez above all other painters and would dearly love to go to Madrid to see the originals of those works. We stayed overnight with the Watsons in their villa in St. John's Wood.

I had not heard for a long time such a hum in the apiary as this afternoon. We had the Duckmantons of Polegate to lunch, and I took them down the garden to hear it. That merry hum is music to my ears, especially as it has been silent these five months. It makes me realise that Springtime is here, which in turn makes me realise that the time is at hand when I too must be up and doing. At the moment my one duty is to see to the food supply, and with as little disturbance to my colonies as possible.

It may be that my successful over-wintering is in part due to my using felt as covering for my hive-roofs, on the kind advice of Mr. J. Brewer of Lewes, whose lantern slide lecture, "Successful Bee-Keepers", I attended last October. I have used a felt smooth, close-made and about the thickness of ordinary shoe-leather, stretched tight over the roof and firmly fastened under the edge all round with flat-headed tacks, and well-painted. The felt remained dry as a bone inside all Winter, and there was absolutely no leakage.

Friday, 16th March I received today the first edition of a new fortnightly bee-keeping paper, *Ptschelorodnaya Zschisu*, sent from St. Petersburg by my friend there, the engineer Vladimir Shukov, he of the Upper Trading Rows in Moscow and the astonishing exhibition pavilions of the All-Russia Exhibition in Nizhny Novgorod. I already receive *Die Europäische Bienenzucht* from Pastor A. Sträuli, of Linz, and *Bee-Keeping in the Argentine Republic* from Señor Adolfo Gomez, of Mendoza. I have, perhaps unwisely, promised the two latter journals a paper on "The Italian Bee".

I have become an object of wonder to our postman. He has never seen parcels, packages and letters from such a variety of countries as he brings to my door. But he is pleased with me; I give his daughter the stamps for her stamp album.

In the apiary, the last lot of fondant has been put on those of my hives that needed it, which should see the bees fed for the next two or three weeks. After that I should be all right feeding them syrup, as the bees will be able to get out and fly to relieve themselves.

Monday, 19th March At last – twelve hours sunshine in the day. For about ten days, at the end of February and onward, the temperature was high and the weather abnormally fine. This was followed, in the best English weather tradition, by a sharp snowstorm on March 8th.

Bees flying freely, working on the crocuses and their golden pollen. The process of loading-up is an interesting one. The young bees especially roll themselves in gay abandon over the anthers, and rub up against style and stigma until they become veritable dusty millers.

Wednesday, 21st March Owing to the hard Winter vegetation is more backward here than for two years past, but the clover leys are promising, and the sainfoin (Onobrychis viciifolia) fields have not been stripped by myriads of wild pigeons as last year.

Lord Avebury tells me he plays his violin to the bees. I did the same this morning. They showed no sign of pleasure, or any other emotion. Their composure and gravity was a splendid rebuke to my impertinence.

Friday, 23rd March Sowed under glass tomato, balsams and petunia.

Friday, 30th March The present month has been a deplorable time for the poor bees. The first week it rained nearly every day, the second was dull and cheerless, and not till the 17th did we have a

Spring-like day, since when there have been frequent falls of snow, with winds as keen as if Sussex had been transferred to the North Pole. I have ensured that a supply of artificial pollen is available in a sheltered spot near my hives to tide the bees over against the barrenness and lateness of the season.

This morning I sat and listened to Fiona and Mlle. Zabiellska play a piano duet together – M. Debussy's *Petite Suite*.

Tuesday, 3rd April With the advent of April a spell of new life has been infused into the apiary by the three days of brilliant sunshine we have been blessed with. The early fruit trees are again putting forth their snowy blossoms.

I was moved on Monday to see the graceful movement of the willows' green tresses by the stream at Monks Farm. I recalled the willows in the Kyi Chu Valley, near Lhasa, which I saw some years ago. They moved in just the same way. In Lhasa itself there is a walled enclosure in front of the Jokhang Temple which contains the stumps of willows known as the "Jowo Utra" (Hair of the Jowo), which according to tradition were planted by Queen Wen Ching at the time the temple was consecrated, twelve hundred years ago.

Wednesday, 11th April A fortnight of fine weather has been followed by snow, sleet, and cold showers, just enough to germinate lately-sown seed and revive the grass crops, and start white clover and vetches into growth. The bees have put in a fortnight of busy days on coltsfoot and willow catkins. Slight night frosts have been prevalent, but the daytime has been of June-like warmth. On Sunday the horse-chestnuts were already showing flowers, and all these signs of advance in nature mean early brood-rearing, with enlarged brood-nests. Should we get another spell of cold weather I must take care to avoid scarcity of food, or the results may be disastrous to the stock.

Thursday, 12th April Sad news that last Saturday Mount Vesuvius erupted and devastated the town of Naples, with lava fountains and violent explosions, and ash clouds reaching ten thousand feet into the air.

Lawn rolled for the first time, which is good exercise, and planted potatoes.

Sunday, 15th April (Easter Sunday) To Telscombe to lunch with the Gorhams. Ambrose Gorham showed us an Easter egg he had had made for his children, four feet high, made of artificial roses and studded with tiny electric lights. His eldest daughter has a pony bedecked with purple ribbons as her Easter present.

Wednesday, 25th April The bees have had a fine day today among our fruit trees, which are a mass of bloom, as are also the wild cherry trees in the woods. The fields and hedgerows between here and Polegate are following suit. Bee pests are still with us. I killed six queen wasps found in my hive roofs a week ago. The previous fortnight, being warm, had roused them from their winter torpor; then the colder weather had, I suppose, driven them to shelter wherever they could find it, in the dry, warm roofs of stocked hives.

Thursday, 26th April Pierre Curie has been killed in a road accident. It happened last Thursday. The news was known immediately round the world, but I have neglected the newspapers these last few days. Walking across the Rue Dauphine in heavy rain, he was struck by a horse-drawn vehicle of six tons in weight, carrying military uniforms, and fell under its wheels, one of which fractured his skull. Marie is prostrated. We shall attend the funeral.

News also of an appalling earthquake in San Francisco on the 18th, caused by movement of the San Andreas Fault. It has destroyed much of the city and killed at least 3,000, with 300,000 left home-less.

Friday, 27th April Today I had an interesting experience with a worker bee. There were a few bees outside the hive, stiff and un-moving from the cold. I picked them up. These encounters pro-vide an opportunity for close observation. There were ten of them, along with a drone. I held them cupped in my palms. One worker stands out in memory. She had a piece of pollen stuck to her wing. The weight was a problem, and it was preventing her wings from locking together properly so she could fly. It would surely cause her death. I cupped the bees in my hand, leaving a small portal open between my thumbs. Several were moving slightly when I picked them up. Some warmed up quickly and popped out of the portal between my thumbs, one after the other. The rest were colder and needed more time. I could feel them crawling around in my cupped hands. Then one by one they too flew off, the drone first, followed by the other workers. The worker with the pollen on her wing flew too, or at least tried to, but she could not, and she ended up on the ground again. I picked her up and took her to the house, where I found a knife I thought I could use to scrape the wing while she sat on my hand. She let me lift her wings so I had better access. I had no luck, but I could now see that the pollen was stuck beneath her secondary wing. I realised she might be hungry, and gave her some honey. While she licked at it I used my dampened handkerchief to lift and rub the wing. I was astounded when she actually lifted and separated her wings, holding them up so that I could swab them over and over again. She never once tried to fly away, and she showed no sign of anxiety. When the wing was clean she flew, landing immediately on the window, as if tired. I offered more honey on my finger. She crawled onto it, her long tongue coming out. I took her outside and stood in a sunny spot in the garden. She ate more honey for a while and then walked across my palm and lifted up in flight. She circled me several times, making orientation loops, and then she was gone.

The pollen on her wing was white, from which I take it she had been visiting the nearby hawthorn. From this year's observations I have been able to compile the following list of the colours of the different pollens and their origins:

Lime – *bright green*
Clover – *brown*
Hawthorn – *white*
Willow Herb – *slate-grey*
Raspberry – *off-white*
Wild Cherry – *white*
Sycamore – *green*
Gorse – *orange*
Broom – *yellow*
Horse Chestnut – *pink*

Monday, 30th April We are now at the end of April and most of it has been miserable for the bees – far too much rain and not more than a dozen sunny hours.

The bees are well fed with plenty of honey left in the supers, just in case we had this sort of Spring. I also added some fondant, to give me peace of mind.

In spite of the rain, the temperature is now higher than we expect at this time of year. However, the amount of pollen the bees have in store causes me concern. With low amounts of pollen the nurse bees cannot produce the bee milk necessary to feed the brood. This means the queen will reduce the number of eggs she lays and the stock will not expand as I would hope.

Tuesday, 3rd May May has come, and with it the cuckoo. Temperatures still high, but with no increase of sunshine, and consequently little bee-work. This month bees will be collecting nectar and pollen from flowering currant, dandelion, willow, cherry, gorse and blackthorn, leading, I hope, to rapid colony build-up.

I have taken to getting up early twice a week, to take a solitary walk and enjoy the dew on the Downs.

Monday, 9th May Rough and cold, more like early March than genial May, and the cold accompanied by wet, windy days, retarding the progress of bee-life and delaying the build-up of stocks.

We leave for Paris tomorrow.

Tuesday, 15th May We are now back from Curie's funeral, which was very well conducted.

There is only one bright spot in this tragic affair. The department of Physics of the University of Paris has decided to retain the chair that was created for Curie, and to offer it to Marie. She intends to accept it, hoping to create a laboratory of international distinction as a tribute to her husband. She will be the first woman in eight hundred years to become a professor at the University of Paris.

Thursday, 24th May The weather is the bee-man's chief topic when he meets his brother craftsman, and as we are now in the fourth week of May, with very few real bee days during the Spring so far, one has not much cheering news to pass on. So far as my own little apiary is concerned, my stocks are backward. Last year at this time I was dealing with swarms and putting on supers, but this year I am employed in feeding my bees. I expect to lose bees through their getting chilled when out foraging in the bleak fields. The colonies are basically strong and would – given a week's real bee-weather – soon render a good account of themselves, but there has not been proper rain in this district for nearly six weeks; consequently, vegetation is backward, with every prospect of a light honey-crop.

Today was Ascension Day, and the children in the village school had a half-holiday. There were races on the vicarage lawn, with iced buns and lemonade.

Sunday, 3rd June News has arrived that last week Princess Victoria Eugénie of Battenberg, grand-daughter of Victoria R, and King Alfonso XIII of Spain, were nearly assassinated when the Catalan anarchist Mateu Morral Roca threw a bomb at their carriage as they returned to the palace from their wedding in Madrid. The bomb was wrapped in a bouquet of flowers. Fifteen people were killed and many more injured, and the bride's wedding gown spattered with blood.

Jas. Hiams told me this old Sussex rhyme when I told him of the mayhem in Madrid. It is to be said to the bees on the death of their owner:

> "Wake, little brownies, wake.
> Your master's dead.
> Another you must take."

He also told me the story of the funeral of a woman his father knew who had kept bees for years. When her body was being lowered into the ground, a swarm of bees issued from one of the hives near at hand, which she had tended, and clustered on a bush near her grave. The swarm settled quietly, and did not disturb the service.

Thursday, 14th June To the Crystal Palace ground yesterday to see a W. G. Grace XI in their match against the West Indians. The

94

tourists needed 376 to win but never got near it, all out for a miserable 128. Bertie Harragin of Trinidad hit three sixes in one over off the bowling of the good Doctor, which amused the crowd.

Thursday, 21st June A few hours of sunshine on the 16th heralded a heavy downpour, followed by a cold, sunless day yesterday. Today half a gale is blowing from the west. Nevertheless, with my hat tied onto my head, I examined the roses for insect pests, and syringed for aphids.

Saturday, 23rd June It has been a full week of dull, cold NE winds, in the middle of June! But as I write I am cheered by the beginnings of bright sunshine and the barometer in the hall rising towards "Set Fair". There is a soft breeze from the SW, and the bees are pouring from the hives.

Sunday, 24th June Today we said goodbye to the five young Jamaicans who have been staying with us since Tuesday. They are students at University College, London, and members of the Kingston and St. Andrew Young Bee-Keepers' Association. I took this photograph in the garden this morning. At the back are Tom Finlayson, on the left, and Eldridge Brown; in the front are, from left to right, Dorcas Powell, Virtue Clarke and Stacia Finlayson.

Monday, 25th June We have had a weekend visit from Watson. Rekha is at a women doctors' conference in Guildford. I took him to Black Cap, along the ridge of the Downs, just west of Lewes. We saw a White Admiral, for me the handsomest of the tribe, its white panels shining against the black. It should not have been in that exposed place, being a forest butterfly. We also saw a buzzard performing the most astonishing acrobatics, rising high up in the sky, turning and plummeting downward in a spiral, and then rising immediately upward to repeat the exercise.

We were caught in a thunderstorm on the way back and sheltered under some thorn bushes by the path. While we were hunkered down there comfortably enough, and sharing the contents of my hip flask, Watson asked me if I was ready to confide in him over the matter of the Whitechapel murders of '88. I told him I was indeed consulted, first by Charles Warren, the Commissioner, and then by James Monro. Then I heard from George Lusk, of the so-called Whitechapel Vigilance Committee, about the "From Hell" letter, which came to Lusk postmarked 15th October, accompanied by half a kidney. I still have that letter.

I told Watson the whole story. I examined the bodies – a grisly business – and spent many days and nights that October walking the ground in the guise of a Portsmouth sailor off one of the cattle boats, and sitting in public houses buying drinks for any who would talk to me. My tattooist friend, Gartrell, who lives in Rotherhithe, gave me temporary tattoos on the backs of my hands – "All's Fair", on the right, and "in Love and War" on the left – and an anchor on my right forearm. I reached my conclusion in the first week of November, and told Warren who the murderer was – the Irish-American mountebank, Francis Tumblety, who was living in Whitechapel. In the end it was not my painstaking hours drinking and listening that led me to him, but the handwriting and the spelling in the letter, and the researches on my behalf by my fellow-detective, Clem Burrows, in Washington D.C. Burrows told me Tumblety kept a collection of female internal organs in his house there, and used frequently to burst out into rants against prostitutes when drunk. On my evidence Warren had him arrested two days later. It was a trumped-up charge of "indecency", so as not to alert the newspapers, but Warren then went too far in that direction and allowed him bail, an act of the most astonishing shortsightedness. The next night poor Mary Jane Kelly was hacked almost to pieces, with Tumblety nowhere to be found, and then some days later news came he was in France. Within a week he was back in the United States, and out of reach.

Why did you not share the investigation with me? asked Watson. Forgive me for reminding you abruptly of your dear Mary, I said, but at that time you were only six months married. The new Paddington practice needed your undivided attention. And, of course, it was necessary for me to work alone in Whitechapel. You have gifts, Watson, but no gift for disguise. Your name was never linked to the investigations, he said. I told him I had made my anonymity a condition of my looking into the case. Then the rain eased, and we had a pleasant walk home and an excellent pie of some description, washed down with Jas. Hiams's robust cider.

In writing this, I am reminded of another case of disastrous police indecisiveness, some two years before that investigation. I have checked the entry in my scrapbook. Frederick Abberline, who was Chief Inspector then, was very slow to execute a warrant in the Cleveland Street brothel affair. The delay allowed the brothel-owner, whose name was Hammond, to get away to France, disguised as a priest. Abberline also failed to act swiftly in the matter of the Hundred Guineas in Portland Place, another transvestite club where the Duke of Clarence was a member.

Wednesday, 27th June This morning on one of my hives, on the alighting board, I saw hard, grey pollen pellets. The pollen has gone stale and hard and has been thrown out by the worker bees. If crushed between the fingers the pellets break up and show layers, and sometimes a trace of colour is visible. The appearance of the pellets is a good indication that the bees are expanding their brood nest.

I have also observed Chalk Brood mummies at the hive entrance. Spores of a certain fungus are present on the bees, comb and hive parts. A drop in temperature combined with high carbon dioxide levels allows those spores to germinate, and it is likely that a protein deficiency in the bees allows them to grow. After germination the vegetative growths of the fungus invade the larval tissues in the bee cells and kill them after they have been capped. The dead larvae usually become chalky white, and swell to fit the cell. Some become dark blue, or even black. They then shrink and harden to become 'mummies', which the bees must then remove from the hive.

This afternoon there were boys playing cricket on the bit of green in the village called The Fridays, which I understand has that odd name because of the Saxon, Freya, the goddess of love and good luck. Clearly, when St. Wilfrid came this way in 680 A.D., preaching Christianity, the locals stuck with the familiar name.

Friday, 29th June Last Monday the actress Evelyn Nesbit, 22, and her husband, Harry Kendall Thaw, attended a performance of *Mam'zelle Champagne* in New York. In the rooftop garden of Madison Square Garden, Thaw approached the table of Stanford White and shot him three times in the face, killing him. White, an architect, was known to have seduced Nesbit when she was sixteen. Thaw, a possessive husband and an unbalanced mind, killed White in an act of jealousy and drunkenness. There is no interesting feature whatever in the case, except in comparison with the murder of his wife, in May this year, by Albert Lemaître, the sporting motorist, in Paris. When his wife told him she loved another man and wanted a divorce, Lemaître shot her twice, then shot himself in the head, but he has not died, and is expected to recover. *Le Figaro* speculates that Lemaître will receive leniency because the act is *crime* *passionel.* One wonders if an American jury will take a similar view of Thaw's action.

Wednesday, 4th July Field-beans are in splendid bloom, and honey is plentiful. Two of my hives are each supered with crates of shallow-frames, and the frames in each hive are crowded with bees.

Two gipsy women came to my garden gate this morning and tried to sell me a nanny goat for 15/-. It was a good-looking goat, as goats go, and think I missed a bargain, but we have no need for one.

Saturday, 7th July I have had a letter from my friend and correspondent in Vienna, Arnold Schoenberg. He is a most exciting composer, only just out of his twenties, that I met through his tutor, Alexander Zemlinsky, who came to know of me through my paper on the motets of Lassus in the 1890s, and has corresponded with me since then. Schoenberg and his wife Mathilde have just had a second child, a boy, Georg.

Chrysanthemums fed and thinned.

Tuesday, 10th July Very odd swarming behaviour is happening this year in my area. Swarms are emerging before the queen cells are fully formed, there is more than one swarm emerging at the same time, and swarms are emerging in the rain. I can only put it down to the very odd weather patterns we are experiencing this year.

Friday, 19th July We had St. Swithin's Day on 15th July and the bees are now in the last period of honey gathering from the last major source of the year, the limes, along with what is left of the white clover. There is a good show of blossom on the trees, but the weather is dull and cool, with westerly breezes. One longs for a week's sunshine with the thermometer at eighty degrees in the shade. It makes me smile to think for how many years in London I paid virtually no attention whatever to the weather, unless it was a factor in a case. That urban indifference has been transformed into something approaching an obsession, as with all my fellow bee-keepers.

Sunday, 28th July A walk this afternoon in Ashdown Forest with Elizabeth Millner, a friend of Fiona's from Baker Street days. In this forest is evidence of human occupation dating back to 50,000 years ago.

Friday, 2nd August A grand spell of fine weather has turned my recent wails into a song of joy. Jas. Hiams told me the other evening that his strongest hives had "a gallon of bees" hanging outside the hives every night, and that they were making honey fast. I am thankful to say my bees have been doing the same, early and late, and the racks of sections I put in on July 19th are nearly completed. This is from various sources, such as limes, white clover, and blackberry.

Sunday, 4th August There have been five cases of scarlet fever in Eastbourne, all notified in the last week. This is very serious. Our friend Dr. Ground says the hospital will use a serum developed in horses which has shown good results in London.

Thinned out dahlias and staked them. Fiona removed top shoots of chrysanthemums to put in small pots as cuttings to make dwarf plants.

Friday, 9th August Yesterday a hive of bees caused a wild commotion at Burgess Hill Railway Station. A large parcel addressed to a local resident was tumbled out onto the platform by a railway employee. The parcel, which contained a hive, gave way, and out streamed the bees, causing the porter to run for his life. The stationmaster ordered, threatened, offered rewards, but none of the porters was daring enough to approach the parcel. Finally, an outside porter, an elderly man, was induced to lift the parcel by the offer of a shilling, but when the bees buzzed round his head he dropped the package and fled. All day the bees held the staff and passengers at bay, but at eight o'clock at night they returned to the hive, which was then hurriedly closed up and carried to its destination.

Thursday, 22nd August The King has gone to see Emperor Wilhelm to discuss the developing rivalry between the naval forces of the two nations.

I was summoned for jury service two weeks ago. I wrote back saying who I am, and how likely it is that I might know some of the men standing in the dock, and some of the policemen giving evidence from the witness box. Today I learn that I have been discharged.

Saturday, 24th August This late afternoon I walked by myself over the Downs for a couple of hours, in the Warren Hill direction, and then lay down to rest. I put my hands behind my head and looked up at the cumulus clouds riding like an armada in a sky of the purest blue, and thought of them stretching away over Surrey, and London, and over the Midlands, and my birthplace. The intimation came to me, not at all sadly, that I shall probably end my days here. As I walked home the sun went down like a blob of molten gold among grey clouds at the horizon, and lit up the western sky with pink and yellow streaks.

Sunday, 25th August Yesterday afternoon we were picking limpets at low tide with Sidney Bolton, Catriona Jarvis and Catriona's sister, Clarissa. We are to eat the limpets tomorrow. Bolton has a recipe that involves boiling and slicing them, then adding lime juice and Chile pepper.

Mr. Knewstubb, a visiting clergyman, preached a sermon this morning on the text from Proverbs: "My son, eat thou honey, because it is good, and the honey-comb, which is sweet to thy taste; so shall the knowledge of wisdom be sweet to thy soul."

Wednesday, 5th September After more than nine weeks of indifferent weather we have had hot sunshine for a week or two, which has excited the bees, and caused them to start brood-rearing in earnest.

After the limpets in August, the blackberries in September. We spent yesterday afternoon with the Hiams's on the edges of Friston Forest, gathering basketsful. Jas. Hiams says they must be got before the end of September, as after that the Devil breathes on them, and they go sour. Fiona plans a blackberry sauce for venison, Mrs. Hiams a blackberry filling for pancakes, and Jas. Hiams will mash them with sugar, sieve them into a bottle and add gin.

Friday, 21st September Michaelmas daisies everywhere! Mowing of the lawn, storing of turnips, asparagus beds thoroughly watered.

I spent yesterday in the village school, at the invitation of Mr. Edward Drew, the schoolmaster. I was asked to talk to classes about honey bees and beekeeping. I took an empty hive and all my bee clothing – veils, helmets and jacket. I also took a smoker and some sample honeycombs, containers with tiny wax flakes from wax glands and a container of pollen. The children enjoyed putting on the bee gear, but what they enjoyed most was handling the beeswax. Their little fingers just needed to test its texture, and so they squished it. I also handed around a paper wasp's nest, and they mashed that, too. Some of the older ones will be coming to New House Farm in the Spring when the bees fly again.

Most of the pupils reach Standard IV or V, and they seem happy to be in school. Their morning yesterday consisted of: prayers by the minister, Mr. Evans; hymns, with Mrs. Drew on the portable harmonium; scripture and arithmetic; a break outside; reading, writing and poetry recitation; then, lunchtime. After lunch they had me, but usually the afternoon is for history and geography, singing, gardening, needlework (aprons and pillowcases, as far as I could see), carpentry, P.E., and occasional Nature Walks. My private education was nothing like so full and varied. Fees are two shillings a term.

Saturday, 22nd September This photograph of mine, of which I am proud, shows four generations of the Deeprose family who farm the Michel Dene Farm near here, with three of the daughters of the family and two Hindoostani men (at left, in the foreground) who come to help with the harvest each year. They were originally lascar seamen who came ashore at Newhaven some years ago and decided to stay. The imposing figure in the centre is Angel Deeprose, aged 89, and legendary for his strength in the fields. Also shown are the horses, Amber and Colly, and the now-retired sheepdog, Blue. I am very grateful to them for taking time off their work to assemble themselves for the photograph.

Wednesday, 10th October In another month or so the bees will go into their winter clusters and then nothing more can be done to help them through the harsh weather until March. The only thing is to make the hives as secure against wind and wet as best I can.

Storage of apples, pears and vegetables with Bill Frusher.

Wednesday, 31st October About a year ago Johann Dzierzon was taken ill, and the worst feared, but his vigorous constitution enabled him to rally, and during the Spring he got better. Towards this Autumn, however, his strength began to give way, and he sank rapidly. The end came on October 26th. To bee-keepers his loss is great, and his name will stand for ever in the history of bee-keeping, and of science. He received some hundred honorary memberships and awards from universities, societies and organisations, as well as the Austrian *Order of Franz Joseph*, the Bavarian *Merit Order of St. Michael*, the Hessian *Ludwigsorden*, the Russian *Order of St. Anna*, the Swedish *Order of Vasa*, and the Prussian *Order of the Crown*.

Tuesday, 6th November Yesterday to Lewes with the Scrope Viners and their friend, the writer Fern Bryant, to watch the extraordinary Bonfire Night revels. I had not expected to see seventeen burning crosses being carried. They commemorate seventeen Protestants burned to death outside the Town Hall between July 1555 and June 1557, four of them women.

Thursday, 8th November Heads of tea and other tender roses protected with dried bracken.

Friday, 16th November The end of three days in the Soup Kitchen. I have been successful in getting rid of the demeaning questions as to eligibility that our staff were told they had to ask. Eligibility for soup and a piece of bread! One look at the hundreds we have served this week decides that question.

I ate with a beggar called Tom who gets his living by whistling. He has been blind since birth. He leans on his cane at selected spots up and down the coast, his terrier Jingles and a cap at his feet, and whistles hymns, popular songs, arias from the opera and traditional airs. He whistled "Abide with Me" for me, with some of the staff crying by the end.

Thursday, 22nd November The weather continues mild, with the bees still working and bringing in pollen. So far, we have had no frosts. Dahlias are still in full bloom, but the last bee-forage, the ivy, is over, and with colder weather imminent my bees will soon be confined to winter quarters. A wonderful smell of smoke is in the air, from the burning of leaves.

Wednesday, 28th November Parliament has passed, at long last, a motion declaring the opium trade with China morally indefensible. I hope the result will be that the government of China ceases the production of opium, and that that long established evil will disappear from our European cities in due course.

We have begun playing whist regularly with the Leekes, whose house is in Upper Willingdon. It is a pleasant pastime, but my ability to remember the fall of every card, and who played it, and when, is steadily making me unpopular round the green baize table.

Monday, 17th December On Saturday to the Bechstein Hall in Wigmore Street for a concert of music by Roland Lassus, including the Nine Lamentations, and, to my great pleasure, the motet *Veni in hortum meum. (Veni in hortum meum soror mea sponsa, messui myrram meam cum aromatibus meis; comedi favum meum cum melle meo; bibi vinum meum cum lacte meo; comedite, amici, et bibite, et inebriamini, carissimi.* I am come into my garden, my sister, my spouse: I have gathered my myrrh with my spice; I have eaten my honeycomb with my honey; I have drunk my wine with my milk: eat, O friends; drink, yea, drink abundantly, O beloved. The verses are from Chapter 5 of *The Song of Solomon.*) It baffles me that this, perhaps the most prolific composer of all time, with more than 1500 religious and 800 secular compositions to his name, a master at the age of 25 and known throughout Europe as the "divine Orlando", should have ended his life in ten years of the deepest hypochondriac melancholy.

Protected the globe artichokes by covering the crowns with ashes, and with Bill Frusher made a mushroom bed.

Friday, 23rd December I have just sent off to the French magazine *Le Jardin de la Santé* an short article on the chemistry of sugars, with attention to: saccharin, its usefulness but lack of nutritive value; saccharose, or cane and beet sugar, which imposes severe work on the organism, and does not suit those whose digestive system is less than robust; and the natural glucoses – dextrose, levulose and, of course, honey, the prince of sugars. (Saccharin was first produced by my friend, Constantin Fahlberg, a chemist working on coal tar derivatives in Ira Remsen's laboratory at the Johns Hopkins University in Baltimore, Maryland, USA. In 1880, I was able to make a small contribution to his work on benzoic sulfimide, and again later, from Montpellier, in 1893.)

Tuesday, 27th December Another Christmas away from home, with the Gorhams at Stud Farm in Telscombe, and a crowd of their friends from the racing fraternity. The day noisy, but full of good cheer, and a bracing ride over the Downs on Boxing Day. Telscombe is a curious little village, only reachable at the end of a road which goes nowhere else. Gorham owned and trained Shannon Lass, out of Mazurka by Butterscotch, which won the National at 20 to 1 in 1902. I remember the race well; the going was heavy, the weather clear. On Watson's advice I backed the joint favourite

Drumree, the Duke of Westminster's horse, at 6 to 1, which came in tenth, or twelfth, or somewhere equally unhelpful.

My wife, standing at my side, insists I reproduce here, from the printed menu, the absolutely extraordinary Christmas Day meal devised by Gorham's American wife, Daisy:

Crème Windsor
Oxtail Soup
Barquettes d'Ecrevisses Nantua
Truite Saumonée au Vin de Chambertin
Baron d'Agneau de Pauillac aux Morilles
Canetones de Rouen à l'Archiduc
Sorbets au Kummel
Spooms au Cherry Brandy
Poulardes du Mans Truffées
Foie Gras frais à la Souvaroff
Salade Gauloise
Asperges d'Argenteuil sauce Mousseline
Petit Pois nouveaux à la Française
Timbales de Fruits Glacés à l'Orange
Glace Viviane
Feuillettes aux Amandes
Corbeilles de Fruits

Porto Commandador
Chablis Moutonne
Château Yquem 1874
Château Haut-Brion 1877
Mouton Rothschild 1875
Clos de Vougeot 1870
Moët et Chandon brut Impériale 1889

In the afternoon I found myself on hands and knees on the carpet, involved in building a railway bridge from "Meccano" pieces, a present to one of the Gorham children. This "Meccano" is as excellent a toy as a child can have – nickel-plated steel strips, plates and angled pieces perforated at half-inch intervals and put together with nuts and bolts with 5/32 BSW threads. In fact, it is not a toy. It is an construction system of infinite educational potential for girls – why not? – as well as boys. I want one for myself.

On arriving home this morning, and collapsing on the sofa, we exchanged presents. For her, a black leather handbag in the modern style, with nickel-plated fittings. For me, a Waterman fountain pen and three bottles of bright blue ink, with which I write these words.

Friday, 28th December The Education (Provision of Meals) Act now allows "Local Education Authorities" – I am not sure what these bodies are – to provide free school meals to the poorest children. Bravo! Another blow struck in the continuing war against those grim twins, poverty and crime.

Monday, 31st December Angela Burdett-Coutts died yesterday, aged 92, of bronchitis. We will forego our New Year festivities tonight. She is to be buried in Westminster Abbey, on January 5th, and I imagine we shall be invited to attend. I will not pre-empt the weighty obituaries which are at this moment undoubtedly being penned. I simply say she was an intelligent woman of equable temperament who used her colossal fortune wisely. Like the Duke of Wellington, Michael Faraday and Charles Dickens, I was fond of her, and got on well with her.

Jas. Hiams has tied a piece of black crepe on each of my hives, while singing a doleful song. He firmly believes that bees must be told of important events in their keeper's lives, such as births, deaths, marriages, or departures and returns in the household. This morning, with his hand on my elbow to guide me, I knocked gently on the hives with the door key to my house and told them that Angela Burdett-Coutts was dead. They received the sad news with profound calm. Jas. Hiams also believes that the bees might leave their hives, stop producing honey, or even die, if not told in the proper fashion. He tells me that his grandfather maintained the practice of inviting bees to a wedding or a funeral, on the grounds that they would be offended if not invited.

My wife watched us from an upstairs window, smiling at the little ceremony. She told me later she found it amusing to see the arch-rationalist so fallen from his rationalism. It is easier to be a rationalist in Baker Street than in east Sussex, I said.

As usual at this time of year I include in the journal at this point some information I hope may be useful – in this case, my bee-keeping calendar. I am a little uneasy that there is still not yet enough practical bee-keeping matter in the journal to warrant the title I have in mind for it, if I should decide to publish it – *Practical Handbook of Bee Culture*. But bee-keeping goes on alongside all the other activities of one's life, and is not an exclusive occupation, so why should my journal be exclusively concerned with bees? Anyway, here is something utterly practical – my list of bee-jobs to be done in each month of the year.

January *The Bees.* The queen is surrounded by thousands of her workers in the midst of their winter cluster. Disturb them as little as possible. There is little activity except on a warm day when the workers will take the opportunity to make cleansing flights. There are no drones in the hive, but some worker brood will begin to appear. The bees will consume about 25 lbs. of stored honey this month. *The Beekeeper.* Little work is required from you at the hives. If there is heavy snow, make certain the entrance to the hive is cleared to allow for proper ventilation. If a thaw presents itself (in January or February) you provide supplemental, emergency food for the bees such as fondant (on the top bars) or granulated sugar (on the inner cover). This is time to catch up on reading on bees, and to build and repair equipment for next season. Order package bees (if needed) from a reputable supplier. If stores are required, give a cake of candy. Examine under the roofs for damp, and make any necessary repairs.

February *The Bees.* The queen, still cosy in the cluster, will begin to lay a few more eggs each day. It is still "females only" in the hive. Workers will continue to take cleansing flights on mild days. The bees will again consume about 25 pounds of honey in the month. *The Beekeeper.* Remove dead bees from entrances. Replace any damp quilts with dry ones. Overhaul any hives not in use, and clean and disinfect them. Get your equipment ready for Spring, and tidy up the apiary. Order new sections, wired frames, foundation and other appliances for the coming season.

March *The Bees.* This is the month when colonies can die of starvation. However, if you fed them plenty of sugar syrup in the Autumn this should not happen. With the days growing longer, the queen steadily increases her rate of egg laying. More brood means more food consumed. The drones begin to appear. The bees will continue to consume honey stores. *The Beekeeper.* Early in the month, on a mild day, and when there is no wind and bees are flying, you may take a brief look inside your hive. It is best not to remove the frames; just look under the cover. If you do not see any sealed honey in the top frames, you may have a case of impending starvation. Provide emergency food (fondant or granulated sugar if the cold prevails, syrup if the weather is mild), as fast as the bees can take it. But remember, once you start, you should not stop until they are bringing in their own food supplies. Reduce entrances to one bee space as a protection against robbing. Unite queenless stocks to others with queens. Keep a look out for signs of disease. Sow seeds of bee flowers.

April *The Bees.* The weather improves, and the early blossoms appear. The bees start to bring pollen into the hive. The queen is busily laying eggs, and the population is growing fast. Drones begin to appear. *The Beekeeper.* Continue stimulative feeding, if necessary. On a warm and still day do your first comprehensive inspection. Can you find evidence of the queen? Are there plenty of eggs and brood? Is there a pattern to her egg laying? Later in the month, on a very mild and windless day, you should consider reversing the hive bodies. This will allow for a better distribution of brood, and stimulate the growth of the colony. You can begin to feed the hive medicated syrup. Add warm, dry wraps where needed. Examine stocks for signs of disease or queenlessness. Unite weak stocks, saving the better queen. Prepare equipment for queen rearing. Destroy queen wasps where found.

May *The Bees.* Now the activity really starts. The nectar and pollen will come into the hive thick and fast. The queen will be reaching her greatest rate of egg laying. The hive should be bursting with activity. *The Beekeeper.* Make sure supplies of water are ample. Watch for signs of honey flow and instal supers in good time. Lessen the possibility of early swarming by ensuring adequate room and ventilation. Prepare an empty hive for swarms. Proceed with queen-rearing and prepare nuclei. Inspect the hive weekly.

June *The Bees.* Unswarmed colonies will be boiling with bees. The queen's rate of egg laying may drop this month. This is the period of the main honey flow. *The Beekeeper.* Continue activity as for May. Inspect the hive weekly to make sure it is healthy and the queen is present. Add honey supers as needed. Keep up swarm inspections. Extract honey from combs and return extracted combs to the centre of the brood nest.

July *The Bees.* If the weather is good, the nectar flow may continue this month. On hot and humid nights, you may see a huge curtain of bees cooling themselves on the exterior of the hive. *The Beekeeper.* Continue activity as for May and June, including inspections to assure the health of your colony. In particular, ensure ventilation is sufficient on hot days. Supply young queens from nuclei. Continue to add more honey supers if needed.

August *The Bees.* The colony's growth is diminishing. Drones are still about, but outside activity begins to slow down as the nectar flow slows. *The Beekeeper.* There is no more chance of swarming. Watch for honey-robbing by wasps or other bees. Remove superfluous supers. There is not too much for you to do this month. You may give yourself a short holiday.

September *The Bees.* The drones may begin to disappear this month. The hive population is dropping. The queen's egg laying is dramatically reduced. *The Beekeeper.* Harvest your honey crop. Remember to leave the colony with at least sixty pounds of honey for Winter. Check for the queen's presence. Continue feeding in the second half of the month until the bees will take no more syrup. Unite weak stocks. Reduce the size of entrances again, to prevent robbing.

October *The Bees.* There is little activity from the bees. They are, in the American phrase, "hunkering down" for the Winter. *The Beekeeper.* Continue to watch out for robbing. Configure the hive for Winter, with attention to ventilation and moisture control. Reduce brood nest to space required. Provide warm wraps. Install a mouse guard at the entrance of each hive. Check hives are protected against storms, rain and snow, and set up a wind break, if necessary. Finish winter feeding. Plant crocus and other pollen- and honey-yielding bulbs.

November *The Bees.* Even less activity this month. The cold weather will send them into a cluster. *The Beekeeper.* Attend to winter feeding, if required. Clean thoroughly, and then store your equipment away for the Winter. The main wintering points are: to winter only strong stocks, to provide a sufficient quantity of food, to keep the bees as quiet as possible, to ensure sufficient ventilation, and to avoid damp and the ill effects of storms.

December *The Bees.* The bees are in a tight cluster. *The Bee-keeper.* There is nothing to be done but check on the hives from time to time, that they are watertight and upright, particularly in bad weather.

1907

Monday, 7th January We returned from London this afternoon. The funeral was conducted with a sombre simplicity, as she would have wished. I wore my ribbon of the Order of the Legion of Honour, Chevalier rank, which I was awarded in 1895 in connection with the tracking and arrest of Huret, the Boulevard Assassin.

In the quiet but inclement days since, I have continued with the ordering of my criminal scrapbooks. Ah, Louisa Miles, the greatest confidence trickster of my time! I believe she is still alive. She has a particular genius for the Rented Apartment swindle, which she learned from the Russian, Sophie Bluffstein, whom she surpassed in guile and daring. Miles calls in at a jeweller's, say in Bond Street, and says that she is Miss Constance Browne, secretary to "Lady Campbell." Lady Campbell's exquisite and expensive card shows an address in some wealthy quarter of London. She wishes to buy her daughter a wedding gift of diamonds; could one of the firm's employees bring over some of the firm's finest jewels for her inspection? When the employee arrives, Miss Browne takes the jewels from him and goes into an adjoining room, asking him to sit and wait while Lady Campbell makes her choice. Eventually the poor employee gets up and tries the door, only to find it locked, with Louisa Miles far away by that time, with the jewels. I have no photograph of her in my scrapbook. I wish I had.

Thursday, 10th January The last few days of severe frost have been a "sharp snap" for things apicultural, and a change of temperature will be welcome. We are now coming in to the time of the year when the bees need the food they have stored throughout last year. Up until now all they have needed was just enough to keep themselves alive, but the queen will begin to lay eggs shortly and that will mean grubs to be fed. That is the time when food really is needed. Heft the hives and if you are not sure there is sufficient food, feed them with candy.

Thursday, 17th January With Bill Frusher planted two new pear trees and a cherry.

Sunday, 20th January On Saturday we returned to London, to a box in the Empire Theatre on Mare Street, Hackney, as a late birthday treat for me from my wife. She knows my penchant for the Varieties. I have the bill before me, a three-hour programme, as follows: the Three Mortons; Madame Mona Delza de Vaudeville; Champion Wrestling; Capt. Zeller's Zouave Girls; Miss Ethel Ross-Sedwicke and her Performing Animals; Mr. George Robey as the Merry Mayor; Bella and Bijou; Mr. Harry Lauder; Miss Vesta Tilley, as "The Naughty Boy"; the Poluskis; the van Droysen Sisters; Mr and Mrs Battersby; Marc Riboud and the Harry Rambler Troupe; Mr. Eugen Sandow, the World's Strongest Man; Miss Millie Lindon, The Balloon Girl; the Eight Snowdrops; the Brahmans and their Shadowgraphs; Miss Annie de Montford and her Mesmeric Seance; Mr. George Mozart with Little Albert; the Jandaschewsky Clowns; Dame Thora, the Transvestite Ventriloquist; Happy Fanny Fields; Millie Betra, The Serpent Queen; El Niño Farini; the Bon Ton Burlesquers, and an "Urban Bioscope Treat".

I spoke with Captain Zeller and Dame Thora after the show. Things are not well backstage. The Variety Artistes' Federation is planning industrial action because the London and provincial theatre managers refuse to countenance any increase in wages for the poorest-paid of the performers, stage hands, dressers and saloon staff. Keir Hardie is supporting the Federation, as is Ben Tillett of the Dock, Wharf, Riverside and General Labourers' Union.

The four grand-children of Frank Matcham came on stage at the end of the show and sang a song welcoming in the New Year. Matcham is the architect who designed the Empire. He also did the Coliseum in St.Martin's Lane. Never mind that by this time it was a belated welcome; we enjoyed it.

Wednesday, 23rd January On Monday the theatrical workers at the Holborn Empire finally took strike action. Marie Lloyd was prominent on the picket lines, and is leading the fund-raising activities. She has issued a handbill which says: "We, the stars, can dictate our own terms. We are fighting, not for ourselves, but for the poorer members of the profession, earning thirty shillings to £3 a week. For this they have to do double turns, and now matinées have been added as well. These poor things have been compelled to submit to unfair terms of employment, and I mean to back up the federation in whatever steps are taken." We have sent our own contribution.

Wednesday, 6th February The strike has lasted for almost two weeks, and has ended in arbitration, which promises to satisfy most of the main demands, including a minimum wage and a maximum working week for performers and staff.

I have sent a copy of my draft score of the *Goldberg Variations*, arranged for violin and viola, to Arnold Schoenberg in Vienna.

Friday, 15th February Good weather, so we sowed early peas and beans, cabbage, onions and parsnips.

We are back from remote, windy and bitterly cold – but most beautiful – Dungeness, where we spent two days with the Armatradings in their extraordinary cottage made from three upturned boats. Both husband and wife are artists – she a painter in watercolour, he a sculptor in wood. We walked, beach-combed on the endless shingle, and visited the sheds, now empty, from which Signior Marconi

117

sent the first radio signal to France sixteen years ago. The sea here can be deadly, and the lifeboat is often in use. The Dungeness lifeboat has the unique distinction of being launched by women – The Lady Launchers, as they are called locally. I took their photograph with the Box Brownie.

Wednesday, 15th May I have been unwell since the third week in February. Dr. Wardleworth, my G.P., has been unable to tell me what I have been suffering from. He used the expression "nervous prostration", as Watson did once, but in the last few years I have had no pressure of cases as I did when Watson tried to treat me. Wardleworth brought in a specialist in depression, a man with – in the circumstances – the somewhat unfortunate name of Toombs. He prescribed only rest, and a dreadful diet without meat and alcohol, which I have, of course, ignored. Jas. Hiams and Fiona have attended to the bees during my indisposition.

The writer, Mr. Joseph Conrad, was good enough to visit me on my sickbed. Some years ago, he told me, he spent some months in a hospital in Switzerland with symptoms not so different from my own, including neuralgic pain. He believes the task of writing can sometimes bring on the condition in him. He told me he is finishing a story he calls *The Secret Agent*, which is taking a heavy toll on his health. It sounds a thrilling and utterly modern tale. He has promised to send me a copy when it is published.

He brought with him a very clever, talkative young Welshman, Thomas Lawrence, whose family is known to him. This Lawrence had just returned from a bicycling tour of France, and was keen to meet me. He is on his way to read History at Jesus, Oxford, where, apparently, there is a strong Welsh connection.

I have been able to write a little in the last two weeks, sitting up in bed. I have concentrated my thoughts on the mystery of swarming in bees, and this, still in draft, is the result:

By the end of May the fifty thousand or so occupants of the hive have carried home vast quantities of nectar, and the queen has increased their number by thousands each day. Bees returning from the fields, orchards and gardens begin to loiter at the entrance, hesitating before entering the close-packed mass within. The mysterious instinct to leave their home, with their queen, is beginning to grip them. But one all-important preparation must be made first; those bees who will stay behind to nurse the growing brood must have their own queen to raise the colony to strength again. The workers construct special cells with larger and thicker walls than those around them. The queen deposits her eggs in them in the normal way, but these eggs are supplied with far richer food than the normal, and some days later the royal princesses will emerge as perfect virgin queens.

At this the old queen hurries from comb to comb, stirring up wild excitement among the citizens. Vats of honey are opened, for suspected danger always leads bees to feed themselves against the possibilities of being without food. Then pouring out from the hive entrance they come, a jubilant throng in the very ecstasy, one might think, of extravagant emotion. The queen appears, and flies to a neighbouring tree. They gather round her, and there the swarm will await the findings of the scout bees, who have been looking for some time for suitable new accommodation. The watchful beekeeper will hive them without delay in a new home, before they fly off to settle in some distant chimney or ruined tower, and so be lost to useful purpose.

Two weeks later, in the old hive, the strongest of the princesses is heard piping in her cell. She cuts the capping of her cell from within and forces it open like a round, hinged lid, before stepping out onto the comb. First, she feeds from the nearest honey cell, and then she devotes herself, more often than not, to – massacre. She goes to the other queen cells, tears them open and kills her rivals. Once secure in her authority she inspects the hive and moves about for a day or two apparently unnoticed in the continuing work of the colony. Then she takes to the air and flies in widening circles round the hive. Watching her from the surrounding leaves and flowers are countless drones, each observing her with his magnificent eyes* of some sixteen thousand hexagonal lenses. She flies over them, and instantly they are in pursuit. This, and only this, is what they were born for. The strongest of them eventually overtakes her in flight, and in a brief embrace mates with her, deposits in her twenty five million spermatozoa, falls away, leaving his genital organs in her, and dies. She may sometimes mate with more drones, on other flights. She returns to the hive, to a rapturous welcome from her people, and will not leave the hive again until her own swarming time arrives.

[*Drones, queens and workers alike have two eyes occupying the greater part of each side of the head. These are the compound eyes. They are large in the drone, smaller in the queen and smaller still in the worker. The duties and function of the drone require the very highest development of the eyes, and accordingly we find his eyes extending so that they actually meet on the crown of the head.]

Because of my indisposition I missed the opening of the new Old Bailey building on the 27th February, to which I was invited as a guest of honour by no less than King Edward himself, who performed the opening ceremony. The court has been five years in the building. On the dome above the court, I am told, stands a bronze statue of Lady Justice. She holds a sword in her right hand and the scales of justice in her left. And above the main entrance is inscribed the excellent admonition: "Defend the Children of the Poor & Punish the Wrongdoer". What a change from when I was born! – when hangings were a public spectacle in the street outside the court, the condemned would be led along Dead Man's Walk between Newgate Prison and the court, and riotous crowds would gather to pelt the condemned with rotten fruit and vegetables.

I also received a charming letter from Ulyanov, who asked me to do anything I can for the Russian composer Mr. Nikolai Rimsky-Korsakoff. He has been dismissed from his Professorship of the Conservatory of St.Petersburg because of his championing of the rights of students following the appalling events of 1905 in the city. I have written to Mr. Rimsky-Korsakoff, care of his hotel in Paris.

Monday, 27th May Yesterday our first outing since February, to the Lamb Inn, Worthing, in bright sunshine, courtesy of Captain Minty and his DeDion-Bouton. There was a battered piano in the public bar, and Fiona played *How'd You Like to Spoon with Me?*, *Give my Regards to Leicester Square* and Harry Lauder's *Stop Your Tickling, Jock,* with an encore of *Waiting at the Church*, with the whole bar joining in with "My wife won't let me!" at the end. Back at the Minty's we played croquet on his lawn.

Bill Frusher planted autumn cauliflower and broccoli, and sowed French beans, while I watched from a bath chair.

Monday, 3rd June A swarm issued from my first hive on Thursday last, and disappeared. I fixed a notice to my front garden gate, as follows: "The first boy or girl who gives information where a swarm of my bees have settled shall receive one shilling." Yesterday morning, when I was in church with my wife, a loud knocking was heard at the church door. The minister paused in his sermon. One of the sidesmen opened the door and found there a small boy who had discovered my swarm in his mother's garden, and was determined to be the first to give the information and so secure the shilling. The boy came into the church, genuflected deeply as he crossed the aisle and then pushed himself into the pew close by my side, presumably in case I had thoughts of escape. He had his shilling at the end of the service, and in the afternoon a section of honey, with tea and bread, for him and his two brothers. They showed the greatest interest in my bees, and expressed the hope that they might swarm more often.

May has come and gone, but the proper May weather was, with only one or two exceptions, conspicuous by its absence. Just two days of bright sunshine and genial warmth came to show us what the bees could do if they got the chance. The other twenty nine days were what we generally expect during a bleak March. Fruit blossom appeared and disappeared unvisited by bees, and arabis, wallflower, willow, plane, hawthorn and other early flowers wasted their sweetness, one might say, on the desert air. The farmers' complaints have been loud and long about the want of sunshine in order to begin haymaking.

Tuesday, 11th June I have been able to repay a little of the kindness shown to me by Jas. Hiams in guiding my infant steps in bee-keeping. Mr. Albert Klinckenberg, of Maryville, Missouri, a long-time correspondent of mine in the matter of finger-printing, sent me two pretty spotted Juliana pigs last week, and I have given them to Jas. Hiams. The animals have straight backs, long snouts and longer legs than pot-bellied pigs, and a lean, almost athletic appearance. Apparently they are friendly and outgoing by nature. They will join Jas. Hiams's three sheep, four hens and his cockerel, Copenhagen, named after Wellington's horse.

Poor Fiona is suffering with asthma attacks. We are trying chloroform liniment.

Friday, 14th June I have spent the last two days consulting the *Alphabetical & Trades* section of the *Post Office London County Suburbs Directory*, 1904, at the request of Gregson of the Yard. He needs some help with a case of bigamy. I commend this astounding volume, and its companions, to anyone wishing to get some feeling for the bewildering variety of the inhabitants and occupations of our capital city. It is absolutely invaluable for detective work; how I wish I had had access to it in Baker Street days! I reproduce here the entries where I found the vital clue to the bigamist's whereabouts:

Amlot, Richard, householder, 8 Beechfield road, Catford SE
Amor, John, builder & decorator, 76 Archway road, Upper Holloway N
Amor, Vesta (Mrs), confectioner, 766 Fulham road SW
Amos, Charles John, sign writer, 40 Holbeach road, Catford SE
Amos, Francisco, tailor, 73 Catford hill, Catford SE
Amos, Francis William, tailor, 1 Perry vale, Forest Hill SE
Amos, Henry William, tobacconist, 22 High street, Hampstead NW
Amos, Walter, Railway Tavern, South End lane, Sydenham SE

Amsden, Albert, boot maker, 16 High street, Stoke Newington N

Amy Waterlow Memorial Home for Training Servants (Sister Elizabeth, lady superior), Beavor lane, Hammersmith W

Anastasio, George, hair dresser, 112 Edmund street, Camberwell SE

Anchor Brush Co. (The), brush manufacturers, 10 Evena road, Peckham SE

Anders, George Lee, electrical engnr., 33 Peak Hill gardens, Sydenham, SE

Anderson & Chapman, upholsterers, 114 Fortune Green road, Hampstead NW

Anderson J. & Co, oilmen, 220 Cambridge road W

Anderson & Co.Ltd, boot repairers, 53 Atlantic road, Brixton SW

Anderson & Co, drapers, 124 to 128 High st, Peckham SE

Anderson & Co. laundry, 30 Langton road, Brixton SW

Anderson, Arthur, credit draper, 12 Tay Bridge road, Lavender Hill SW

Anderson, Elizabeth (Mrs.), laundry, 37 Duncombe rd. Hornsey Rise N

Anderson, Emily M (Mrs.), Girls' School, 77 Bolingbroke grove, Wandsworth Common SW

Tuesday 18th June Ear-splitting shrieks at midnight last night drew us out into the garden. The culprits were two barn owls, who had caught a rabbit on the lawn. We could just make out their strange, pale, heart-shaped faces, like masks, and their black eyes.

Monday, 24th June Three fascinating days at the shoulder of M. Louis Salzmann, who is excavating the Roman fort at Pevensey. Among animal skulls – oxen, goat, and a single cat – and some leather shoes, we found coins of the mid-third to early fourth centuries, and some arrow heads. M. Salzmann has told me of the remains of human settlement that lie in the hills round my village – tumuli, great earthworks, and lines of primitive field culture.

It has been a great pleasure to wander the marshes and reedy meadows behind Pevensey Bay which captivated Samuel Palmer, John Constable and J. M. W. Turner. One afternoon I walked across that lonely, but beautiful, expanse as far as Herstmonceux. On the walk, I found myself passing a little churchyard, and I stopped for a moment. Something about the great clouds of cow parsley around me under ancient plane trees, and the golden privet, dark holm oaks and copper beeches, and the quietness, brought me such a deep and unexpected joy that I decided to record it here.

This afternoon planted pansies, wallflowers and winter stocks.

Monday, 1st July An interesting crime for my notebooks. There was an armed robbery last Tuesday in the city of Tiflis, in Georgia. A bank coach was attacked with bombs and guns, and 340,000 roubles stolen. The attackers are Bolsheviks. I sense the hand of my friend, Ulyanov, somewhere in this.

Swifts in their dozens over the village green this evening.

Thursday, 4th July Showers of hail were common in the closing days of June, and the hill was white with frost on some mornings. We are still wearing winter clothing! Drones, drone-brood and even immature bees were being thrown out of the hives up to the very end of June, showing plainly that supplies had run short. Although my hives went into winter quarters with a superabundance of stores, I found two of them on June 28th were without an ounce of honey. My system of feeding them, hastily resorted to, was of the rough and ready kind. I made the syrup medium thick and poured a cupful over the seams of bees between the frames after driving them down with a gentle puff of smoke.

Thursday, 11th July With the recent soaking of the pastures and now rising temperatures we should have a full crop of white clover.

With Bill Frusher covered fruit trees with netting, as an experiment. No sign of apple moth.

The Leekes are paying us a visit. We took a picnic tea down to Birling Gap and sat on the beach. Alice, the younger Leeke girl, shrimped all afternoon with an enormous net, while her mother watched her from under a parasol, and I fell asleep listening to the crash of the waves and the rattling of the pebbles. I asked George Leeke, who is a geologist, how the horizontal bands of flint got into the chalk cliffs. He said the cliffs themselves were formed some eighty million years ago, from the remains of unimaginable millions of minute planktonic organisms living in the sea which covered this area then, but there were also sponges in that sea, which when buried under later layers of deposition were transformed by chemical action involving silica into flint nodules, often in extraordinarily flamboyant shapes.

Friday, 12th July Great news. W. M. W. Haffkine is cleared of malpractice in the matter of the Indian villagers who died after being given vaccine against plague. For this he was dismissed as Director-in-Chief of the Government Plague Laboratory in Bombay, but now the heavy guns – Ross, Smith, Leith and Flexner – have written to the *Times* confirming the Lister Institute's finding that the cause of the deaths was an unsterilised bottle-cap, and not a defect in the vaccine. Haffkine, a Russian Jew, has overcome many difficulties in achieving his pre-eminence in his field, not least the prejudice against his race in Russia and in Europe, and he has displayed courage of the highest order. He tested his first bubonic plague vaccine on himself. He saved thousands of lives in the cholera out-break in Russia in 1898, and at this date more than four million Indians have been inoculated against plague. He must be re-instated.

The honey-harvest is fast drawing to a close in this district. Last week we had four days of good heat, and the honey came in rapidly. My good Italians filled a standard-depth super in that short period.

But the good time was all too brief, and since then, with dull, cool weather, honey has come in but slowly. Everything points to a crop considerably smaller than was in sight at the corresponding period last year.

This morning a large swarm from one of my hives hovered a long time in the air, and at last seemed on the point of settling on the top of a horse chestnut tree fifty feet from the roadway by my house. I was with Freeman-Thomas, who asked me if I possessed a shotgun. I said no, but my next neighbour-but-one, Edward Gorringe, has one. I went and borrowed it from Gorringe's wife, Edward being in the fields. It is a Bland, a massive single-shot 4-bore with black powder and lead balls of the kind used to stop elephant in India and Africa. Freeman-Thomas fired the gun, with deafening effect, just underneath the bees. The recoil threw him head over heels backwards into a bed of nettles. Immediately the bees began to descend and in a few minutes had settled on a shrub about four feet from the ground. Freeman-Thomas said that the shock has a similar effect on the bees as a thunderclap might be expected to have, and drives all thought of wandering from their minds, making them seek a low shelter at once.

Saturday, 13th July Planted broccoli and saladings in the kitchen garden, and harvested onions, for drying and storing.

Finally sent off my paper on "The Italian Bee" to the Austrian and Argentine journals I promised it to eighteen months ago.

Monday, 15th July A sad notice in *The Times* this morning – the announcement of the death of Edwin Ridley, the young man I met in Ridley Hall in such strange circumstances last year. He is given as the younger of the twin boys, but the falsehood is immaterial now he is gone. We shall send some flowers anonymously.

Tuesday, 16th July To London yesterday evening with Freeman Freeman-Thomas for a party to celebrate the knighthood of W. S. Gilbert "for contributions to drama". A long-overdue honour, as Sullivan was knighted for his contributions to music in 1883. I was almost overwhelmed to meet three great Opera House heroines in one room: Margaret Macintyre, who thrilled Mycroft and me as Valentina in Meyerbeer's Les Huguenots in '92; the Canadian Emma Albani, a magnificent Isolde in her final season at Covent Garden in '96; and the greatest of all divas, Nellie Melba, whom I saw as Mimi with Enrico Caruso in '02. She was resplendent on the arm of Auguste Escoffier.

My wife took this photograph in Trafalgar Square in the afternoon before the party. It shows me, with Freeman Freeman-Thomas in a boater.

Saturday 10th August The workmen today finished erecting my lean-to laboratory on the north side of the house. They also installed my equipment, such as it is, and I hope to begin some long-deferred work on the acetones immediately. Acetone is an organic compound, the simplest of the ketones, and a building block in organic chemistry. To begin with I shall attempt the dry distillation of acetates, for example, calcium acetate in ketonic decarboxylation.

Friday, 16th August Yesterday to Horam, near Heathfield, to the workshop of M. Fourdrinier, a violin maker and repairer, to get some new pegs and to have my purfling repaired. The place is full of exotic woods from all over the globe, and a vast range, of course, of spruce and maple. He was dismayed at the condition of my instrument, and suggested a re-varnish, which began a deeply instructive conversation. I learned that as soon as a new violin is completed "in the white", that is, with all wood carving and glueing completed, it must be exposed to the air and sun for a period of several weeks to take away the whiteness of the freshly revealed wood. The air darkens the wood by oxidation, and as the wood darkens, its visual depth and figure become more pronounced. The next step, before varnishing, is the application of the "ground" coating, done in much the same way as painters of old prepared their panels and canvases for painting. The ground coat not only enhances the wood's beauty by further accentuating its depth and transparency, but also penetrates and strengthens the wood. When the overlaying varnish has worn or chipped away with age – as in my case – the ground coat continues to provide protection from perspiration, moisture, dirt and other deleterious factors.

Tuesday, 3rd September I see the Italian government has had to direct so much of its funding to the restoration of Naples and the surrounding area after the eruption of Vesuvius last year that the Italians have asked to be relieved of the burden of hosting the

Olympic Games in Rome. The Olympic Committee has asked London to step in at short notice.

Wednesday, 4th September At sunset, there being a strong smell of flat-irons in the house, I played the violin for an hour in Friston Wood.

Tuesday, 10th September A full house in Old Home Farm. During the last three days we had staying with us three generations of the Brownlow family, from Spitalfields in east London, en route to their hop picking fortnight at the Guinness Northlands Farm near Bodiam. Bill and Margaret Brownlow are the grandparents, Terence Brownlow their son, with his wife Kate and their son Stephen, aged seven, and Susan Brownlow, Terence's younger sister. Terence is a friend of William Wiggins, the captain of my Baker Street Irregulars, who suggested the family come to us for a taste of the country life. Fiona and I are now determined to brew our own beer.

Friday, 20th September As Autumn advances, the supply of nectar decreases as the plants begin to die. It is necessary for the survival of the colony that a limit be set to the consumption of stores. The drones – always heavy feeders, and for whom there is no longer any use – have become the principal danger. They have had their day of indulgence and sunny idleness, and now the time has come for them to quit the hive. Once they leave, sentries at the entrance will not let them back inside. Others cling to the combs until a multitude of pitiless workers sting them to death and remove the bodies. In the chill of an autumn evening an army of males, once so gay and careless, lies motionless in the grass round the hive.

Monday, 23rd September We have spent the weekend at Petworth as the guest of Charles Wyndham, 3rd Baron Leconfield, who succeeded to that title in 1901. I have Watson to thank for the invitation, I suppose, because Wyndham has been an avid reader of

Watson's stories about my cases, and was interested to meet the man behind them. This is fame, of a kind! I must hope I am not too great a disappointment in the flesh. A servant used my Box Brownie to take this photograph of Charles driving his four-in-hand carriage. My wife is in the feathered black hat directly behind the driver, and I am the man in the top hat behind her, in what seems, now I look at it, a slightly perilous situation.

I should not forget to say that the limewood carvings in Petworth, by Grinling Gibbons, are extraordinarily beautiful.

Much gathering of apples, pears and other fruit. Mrs. Trench has taken many bags for jam-making.

Monday, 30th September Heavy rain all day. We have a good selection of bee-flowers in the garden, but it seems to be the opinion of many that garden flowers make but a small proportion of the weight of honey stored, which they estimate to be gathered mainly from clover and fruit blossom.

Honey from the clover family stands supreme, I believe, both in quality and quantity. The combs are capped white, so the product is of fine appearance, and honey from white, alsike, sweet or other clover is sure of a market at a fair price in almost any season. The white clover is a fine pasture plant, common along roadsides and in pastures everywhere. Alsike, or Swedish, clover resembles the white in some respects, although much larger and better suited for culture as a forage crop. It succeeds on land where red clover will not do well, and when sown with a mixture of other grasses makes an excellent meadow. The sweet, or Bokhara, is a biennial, seeding freely and establishing itself everywhere even under unfavourable conditions. My duty to chemistry compels me to add that sweet clover is a major source of coumarin, a fragrant organic chemical compound in the benzopyrone chemical class, which is a colourless crystalline substance in its standard state, used for its sweet smell in perfumes since 1820, and in pipe tobaccos, though not in mine.

Another principal honey-source is fruit trees, which abound in these parts, including orchards of apple (of dozens of varieties), apricot, cherry, damson, greengage, mulberry, medlar, plum, pear and quince, and in my village alone there are two ancient walls laden with peaches in the Summer. A third source is trees such as limes, willows and poplars. Then, of course, come the wild flowers, with which this part of east Sussex seems particularly blessed. We have, to name but the most common: red poppy, corn marigold, corn camomile, field pansy, scarlet pimpernel, white campion, bladder campion, agrimony, bird's-foot trefoil, cowslip, the common spotted orchid, field scabious, cat's ear, grass vetchling, lady's bedstraw, lesser stitchwort, meadow cranesbill, musk mallow, pepper saxi-frage, rough hawksbit, snake's-head fritillary, sneezewort, toadflax, yarrow, yellow rattle, wild thyme, common dog violet, hairy St. John's wort and nettle-leaved bell flower.

As far as the garden goes, the bee-keeper can grow the following nectar-rich plants, and encourage his neighbours to do likewise. **In Spring**: allium, bluebells, bugle, crocus, daffodil, euphorbia, heather, honesty, grape hyacinth, primroses, viburnum, wallflow-ers, daffodil, currant, forget-me-not, hawthorn, pulmonaria, rose-mary and thrift. **In Summer**: aquilegia vulgaris, bergamot, buddleia, comfrey, dahlia (single-flowered only), evening primrose, foxglove, lavender, poppy (annual and oriental), sunflowers, thyme, borage, lemon balm, verbena, honeysuckle, delphinium, everlasting sweet pea, fennel, hardy geranium, dog rose, potentilla, snapdragon, sta-chys, teasel, thyme, verbascum, angelica, aster, cardoon, corn-flower, sea holly, fuchsia, viper's bugloss, globe thistle, ivy, penste-mon, scabious, sedum and verbena. **In Autumn**: asters, colchicum, Japanese anemones, salvias and the tansy-leaf aster. **In Winter**: clematis, hellebores, mahonia, sarcococca, snowdrops, winter aco-nites, winter heathers and winter honeysuckle.

Bees can apparently see purple more clearly than any other colour, so the purple plants among these should be especially valuable, such as lavender, alliums, buddleia and catmint.

I also now keep in my garden a log pile as nesting for bumblebees, and for solitary bees such as the leaf-cutter. Both are excellent pol-linators. The forensic study of pollen, known as forensic palynol-ogy, and not known to me in my consulting days, can tell the in-formed observer a great deal about where a person or object has been, because regions of a country, or even more particular loca-tions such as a certain set of bushes, will have a distinctive collection of pollen species. Pollen evidence can also reveal the season in which a particular object picked up the pollen, and can even produce specific findings of location of death, decomposition and time of year.

In respect of sources of nectar it may be of interest to share the observations I have made over the last two and a half years of the visitation rate of bees to certain crops. I have evidence that the bees spend ten seconds or so per visit to the apricot flower; sixty seconds to the apple; eighty seconds to the cherry; one hundred to the raspberry; and one hundred and thirty to their favourite – at least in my garden – the blackcurrant.

The scrapbook work goes on, on days like this when outside work is impossible. Dr. Thomas Cream was a Scottish physician, and a murderer. He was also cross-eyed, which was his undoing four years ago. In 1891 and 1892 he poisoned prostitutes in south London by giving them strychnine pills, saying they were medicine good for the complexion. While he was killing these women, he was also writing to the newspapers offering to name "The Lambeth Poisoner" for a reward of £30,000, and complaining to the police on numerous occasions that prostitutes were following him. When Louisa Harvey told the police that a cross-eyed client of hers had offered her pills, it was remembered that Cream had that eye defect, and he was arrested. Bottles of strychnine were found in his lodgings, and he was hanged in November 1892.

Wednesday 16th October We are just returned from France, courtesy of the Dieppe to Newhaven ferry. We went to Paris to see a retrospective exhibition of an extraordinary painter, virtually unknown in this country, M. Paul Cézanne. He died in October last year. There were fifty or so works on show in the established genres of landscape, still life and portraiture, but the paintings themselves are revolutionary. M. Cézanne has put aside traditional elements such as harmonious pictorial arrangement, highly-finished surfaces, single-view perspective and outlines that enclose colour, in favour of sustained experiments in the geometric simplification of visual phenomena. This technical aim is accompanied by what I can only call an emotional immediacy, a raw presentation that by the magic of his use of colours makes an immediate impact on the senses, and – hard though it is to believe – makes the work of many of the old masters now seem formal, over-studied, over-varnished and, no matter what their skill, somewhat theatrical. And he was a hard worker, typically taking one hundred working sessions to achieve a still life – I have never seen such apples in my life! – and perhaps one hundred and fifty to produce a portrait.

I am not at all surprised that artists of the present era should feel a need to create a new kind of art which will encompass the fundamental changes that are taking place in technology, science and philosophy. Indeed, how could they not? Photography, which will render much of the representational function of visual art obsolete, is bound to affect this aspect of painting, as is, perhaps to an even greater extent, this extraordinary infant, the bioscope. It is already clear that in painting, colour and shape themselves, not the depiction of the natural world in anecdotal form, will form the essential characteristics of a new art.

We were also privileged to meet M. Auguste Rodin in his studio, and his secretary, a young Austrian called Herr René Rilke, who is also a poet. He gave me a volume of poems, "Neue Gedichte", which is "New Poetry" in English, but sadly that is more or less as far as my German takes me.

The honey season has been very poor, though what little honey has been gathered is good. Swarms have been scarce, owing to the spring months having been so cold, with frost and snow in April and cold N.E. winds all through May, which caused all stocks to be backward in strength. Even when our orchard was a perfect sheet of blossom I had to feed the bees to keep them going. Then in June and July it began to be very hot, with dry winds day and night, so that the clover was largely scorched, and the other flowers relied on for forage did not seem to secrete much nectar.

Thursday, 17th October Sat for an hour after lunch simply watching the hives. While I was there I saw something run very quickly across the grass, onto the landing platform and to the front of the hive. A half second later it ran off in the direction it came in. I sat motionless. Then it happened again, but this time I was expecting it. It was a shrew. It darted faster than I should have thought possible to the hive entrance, snatched a guard bee in its mouth and then ran off with it.

Cannas, begonias and dahlias stored in the shed, which I hope is frost-proof.

Wednesday, 23rd October Yesterday with the Humberstones, Mrs. Guedalla, Miss Dean-Bisset and Miss Francisca Blaauw to Potter's Museum of Humorous Taxidermy, at Bramber. Mr. Potter made his first attempt at taxidermy at sixteen, when he preserved the body of a pet canary. He went on to stuff hundreds of animals during the middle years of the last century, and to create "dioramas" of the most extraordinary fantasy. His daughter Minnie showed us round. We saw a rats' den being raided by the local police rats – perhaps, said my wife, at the instigation of a rat detective; a village school that features 48 little rabbits busy writing on tiny slates; a Kittens' Tea Party displaying the height of feline etiquette; a guinea pigs' cricket match was in progress in one room, and in another 20 kittens were attending a wedding, wearing morning suits or brocade dresses, with a kitten minister in white surplice.

Wednesday, 30th October I have had a letter asking if I will receive, next Friday at 11 a.m., a visitor from the French diplomatic service, one Raymond Lecomte. He seeks my advice on a confidential matter. Shades of Baker Street! I shall meet him.

Friday, 1st November My French visitor arrived this morning in a squall of rain. I watched him from my window as he descended from his hansom. He had on a jet black Norway Seal coat, and was clutching his grey Homburg hat to his head to stop it flying off in the wind. He is a Frenchman, I said to myself, so why is he dressed in the German style? When Pearl showed him into the sitting room she had taken his coat and hat, and he was resplendent in a superb dark suit, white waistcoat and polka-dot necktie. Our conversation was so remarkable that I have decided to reproduce it in as much detail as I can recall.

'Mr. Holmes,' he said, advancing to shake my hand, 'thank you for seeing me. I am relieved to find you, like myself, informally dressed.'

'I apologise for our Sussex weather,' I said, 'but at this time of year...'

'No apology, I beg you,' he said. 'My current posting is Berlin, and the weather there is – how shall I put it? – just as English as it is here.'

'Is the French ambassador in Berlin at present not M. Jules Cambon?' I said.

'You are well informed,' he said. 'I am his First Secretary. I transferred to Berlin before he did, in July 1905.'

'You are very young for such a post. May I ask where you acquired your excellent English?'

'My father sent me to the English lycée in Paris, and I went on to Oxford. Classics – the perfect training for the diplomatic corps.'

His elegantly patent leather shoe tapped repeatedly on the carpet as he spoke. I waited for him to begin. The rain lashed violently against the windows.

'I must share with you a matter of the most exquisite sensitivity,' he said, 'involving as it does some of the highest-ranking military figures in the *entourage* of Kaiser Wilhelm.'

'I confess myself surprised,' I said, 'in view of recent history, to hear that such figures, undoubtedly Prussian, should have enlisted French help.'

He smiled.

'Sir, they have not enlisted my help. I am embroiled in the matter, and am seeking your aid as much for myself as for them.'

I settled back in my chair.

'I am all ears,' I said.

'I shall not mince words. I am, in private – I am sure you will understand the expression, Mr. Holmes – a votary of Greek love, the love that your own Oscar Wilde described as the love there was between David and Jonathan, such as Plato made the very basis of his philosophy, and such as we find in the sonnets of Michelangelo and Shakespeare. For two years now I have been the friend, hunting companion and – yes – the lover of Philipp, Prince of Eulenburg, who is himself one of the Kaiser's intimate circle.'

At this extraordinary revelation, I could not quite control myself.

'Good God,' I said. Eulenburg as the newspapers show him is heavy-jowled, hearty, lavishly bearded and moustached, uniformed and bemedalled – the very epitome of the masculine.

'Is he not married?' I said. 'With children? I have read newspaper articles...'

'He has been married to a Swedish noblewoman for thirty years. They have six children. I myself am married, with two children to whom I am devoted. But your surprise is understandable. It is well known that, in matters of sex, the English have little or no imagination, and are constantly astonished at the behaviour of other nations. A year ago a newspaper made my relationship with the Prince public. Of course, it was denied. Then matters became more serious, and a faction in the Prussian court began to try to unseat the Prince. Articles appeared in the press claiming that he runs a degenerate clique of powerful homosexual men round the Kaiser – the "united fairies", they are called – rendering German foreign policy effeminate, and incapable of challenging Britain's naval and colonial dominance. The latest of these articles, published two weeks ago, links Prince Eulenburg to Lieutenant General Count von Moltke, adjutant to the Kaiser and military commander of Berlin, in the most explicit terms of homosexual love. Are you about to say "Good God" again?'

'I was,' I said. 'I could hardly have been more surprised if you had named Bismarck, or Gladstone. Moltke himself? Can it be true?'

'It is,' he said. 'There is no jealousy among us. We call each other sisters, and believe we all arose together, in joyous union, from the mysterious Spring of Being. But the naming of someone as senior as Moltke means there must be a libel trial. Mere denial is no longer enough. Moltke has challenged the journalist, a Jew called Harden, to a duel, but the offer of honourable settlement has been declined. The difficulty with a trial is that the defence will almost certainly call as witnesses not only Chancellor von Bülow, but even the Kaiser himself.'

'How could the Kaiser be a witness?' I said. 'Surely he does not know of these... goings-on?'

'The Kaiser and Prince Eulenburg have been lovers for twenty years,' he said. 'They hunt together, they both believe passionately in the occult and other affairs of the spirit world, and both enjoy the company of vigorous military cadets, and young fishermen of the Starnbergsee, where they holiday together. Eulenburg is the only courtier who does not address the Kaiser as "Your Majesty". He calls him "Liebchen", or as we say in French, "Chéri". He is also able to soothe the strange and sadistic moods that grip the Kaiser from time to time. Sometimes the Emperor, at parties, will make senior political and military men dance before him as ballerinas, in pink tutus, or dressed up as French poodles. Eulenburg does his best to moderate these excesses.'

'I am sorry,' I said. 'I need a drink.'

I rang for Pearl, and she brought whisky and water for us both. I contemplated for a moment the possibility that M. Lecomte was unhinged, and that this was all some bizarre charade, but every aspect of his demeanour convinced me he was serious. Pearl built up the fire for us, and withdrew.

'So,' I said, 'there will be a trial. Is that why have you come to me?'

'I have come to you, Mr. Holmes, because recently a package of letters was stolen from me – letters from Prince Eulenburg which

140

are my most precious possessions, but which, if produced in court, would establish the allegations against him, and undoubtedly ruin both of us.'

'From where were these letters stolen?'

'From a desk drawer in my study at home, which is just off the Pariserplatz. They disappeared last Tuesday, probably in the early afternoon. I discovered the theft when I returned from the Embassy at five, and saw the drawer had been forced. Since then I have been in a torment of horror at the damage those letters may do, which is why I am here today. I know of your astounding reputation.'

'Who was in the house at the time? You have servants?'

'We have four. Two had the afternoon off. Our cook was in the kitchen downstairs, and the maid was sorting out laundry upstairs. Neither heard anything. My wife was at a friend's house.'

'Was anything else taken?'

'Nothing. But several other drawers had been opened.'

'Clearly, someone who knows the import of the letters has taken them, or, rather, has arranged for them to be taken. It is obviously the work of an accomplished thief, not your journalist enemy, to have entered the house, done his work and got out again like that. The hallmarks of the professional in this area are speed, silence and invisibility. Did you inform the police?'

'I did not. That would begin the unravelling of the secret which I must protect.'

'And no one has been in touch with you in connection with the letters?'

'For blackmail, you mean? No.'

I reflected, and for a few moments we both heard the wind howling round the house.

'I fear that you, and the Prince, must brace yourselves for the consequences. The letters are presumably already in the hands of your enemies. I cannot help you.'

'I am in a desperate position, Mr. Holmes. Allow me, if you will, to put two arguments to you which may change your mind. You have heard of the Tirpitz Plan? No? The Tirpitz Plan is the plan of an influential group in the Kaiser's court who wish to see Germany build up its fleet and land army to a colossal size, and then provoke war with Britain, both in Europe and in the colonies – a trial of strength which would release the destruction of Armageddon in the world. Prince Eulenburg and Count von Moltke advise the Kaiser to follow a less warlike plan – to concede naval supremacy to Britain but to build up the army, and then unite continental Europe behind German military and economic leadership. If my two friends are disgraced, and expelled from the court, the Tirpitz faction will carry the day, and there will be a devastating war in Europe within a decade.'

'I am sorry. I would, of course, do anything in my power to help avert such a cataclysm. Please understand that I am refusing simply because I cannot see how my investigation could succeed, in a foreign city where I know no one, and whose language I do not speak.'

As I said these words a thought came into my mind, that actually I *did* know someone in Berlin – someone who might be very helpful – but I said nothing.

'My second argument is more personal,' he said. 'I am deeply embarrassed to introduce it, but if it is humanly possible to recover those letters, I must stop at nothing. Your brother, Mr. Mycroft Holmes, is mentioned more than once in them. He and his companion, Mr. Crawford, are part of the Prince's Lichtenberg Circle, and have from time to time entertained its members in the most lavish fashion, when they have come to London.'

This gave me pause.

'A hit,' I said. 'A very palpable, blackmailing hit. You know my brother?'

'He has spoken to me only twice. Our longest conversation was in '99, when he questioned me at length about the aftermath of the Dreyfus Affair.'

I walked him to his waiting cab holding my umbrella over his head, ascertained where he was staying – the Grand, in Brighton – and then called my own cab to take me to the Post Office in Polegate.

Sunday, 3rd November The thought which had occurred to me last Friday, and which I did not voice, concerned one Konstantin Eugenides, a Greek with a Roumanian father, and one of Europe's leading cat burglars. He has lived in Berlin for some years, and I am occasionally in touch with him. He sent us charming wedding presents, I remember. He is getting on in years now but still works, in partnership with his eldest daughter, Dora. In the field of robbery without violence, the combination of his experience and skill, and her daring and athleticism, is formidable. He knows me well, and owes me more than one favour, principally from the time – was it four years ago? – when he and Dora were so bold as to rob Abdul

Hamid, Sultan of Turkey, and I had to save their necks. Konstantin should have known better than to steal from one whose nickname is Abdul the Damned.

On Friday evening, thanks to our gallant Post Mistress, I got special access to Polegate Post Office, which was of course by then closed, and sent several telegrams to Berlin, and received several in return.

Monday, 4th November I met with Lecomte this morning in Brighton, and agreed terms with him. Money is clearly no problem in this case. Konstantin's fee is handsome – to be paid in person when Lecomte returns to Berlin – and so is mine. We shall meet again on Wednesday, when I hope to have a surprise for Lecomte.

Tuesday, 5th November A joyful evening in the village. We ate sausages and mash and drank pale ale, lit by fireworks and a bonfire on the green.

Wednesday, 6th November I met Lecomte again this afternoon, in his suite in the Grand. In the highest good humour, I gave him a parcel wrapped in stout brown paper, and he, in tears of joy, gave me a banker's draft, which means the Soup Kitchen in Eastbourne can have a new twenty-year lease on the building. Lecomte, when he was not embracing me, said I must be in league with the devil to have pulled this off. The truth, as ever, is more prosaic, yet more amusing. I did not tell him how the trick was done, but if he ever reads this – which is most unlikely – he will know. Eugenides not only knew about the Berlin theft – I expected that much – but he was the very thief hired by the newspaper to steal the letters! When he received my first telegram he still had them in his possession, because of a dispute with his client about the price for the job. After we had agreed our own price, he put the package in the hands of a special courier, who delivered it to me late last night. It was a dramatic moment, worthy of a novel by Mr. Thomas Hardy: a sudden knock at the door, a slender figure in uniform who saluted me against a background of firework lights in the sky, a few words in German, a signature on a pad, a parcel handed to me in a leather case with impressive straps, another salute – with a click of the heels this time – and he was gone into the darkness. I spent the night reading those letters, I say without shame. Needless to say, the parcel I gave Lecomte was lighter – by the four letters in which my brother's name was mentioned – than the one I received from the courier.

Have I contributed to the peace of Europe, and the Empire, by slowing the expansion of the German navy? Or have I merely saved the skin of the dyed-in-the-wool rogue and anti-Semite, Eulenburg? Time will tell. Fiona shall have a drop-head, high-arm, positive four-motion feed, self-threading, automatic bobbin-winding, patent dress-guarded, seven-drawer Burdick sewing machine, in a solid oak cabinet, from my ill-gotten gains, and I shall drink champagne with her tonight in the Tiger.

Wednesday, 13th November Planted liliums.

Tuesday, 19th November A terrible storm is pelting the Downs with rain, and a fierce wind is roaring in the trees. God help all sailors in the Channel, and all those who have no shelter or warmth. This afternoon we were huddled round the fire in the downstairs sitting room, Fiona reading, and I writing letters. The one I have just finished is to a young man, Herr Kafka by name, who wrote to me last week, in excellent English, on an interesting subject. He is about to start work in the Prague office of the *Assicurazioni Generali*, an Italian insurance company, and, knowing my name from a volume of my case histories, hopes I might be able to give him advice on distinguishing genuinely sick employees from malingerers. I have done my best in a rather longer letter than I intended to write. In a moment I shall put on my waterproof and boots and check that the hives are still upright on their stands.

I hear by letter from Scotland Yard that young Arthur Harding has taken over the leadership of the Walker gang, now that "One-Eyed" Charlie Walker is dead. I wonder if the Titanic Mob are still disputing Brick Lane and the surrounding territory with the Walkers? The newspapers have nothing. I shall write to Harding care of my old Spitalfields informer, Yaakov Puppa.

Wednesday, 27th November Last week I watched for a while a great bank of ivy on the side of Seth Smith's barn, which is down the lane a little way from the back gate of my garden. In the winter sunshine the ivy was humming with activity. The bees were coming out, with fully-laden pollen baskets, at the rate of fifty a minute. On Monday they were still quite busy, but the night of the 25th was a hard frost, and yesterday morning was a short but piercing blizzard. Those last, gallant workers are asleep at last, and will do no more honey-gathering till the aconite blooms next Spring.

Tuesday, 10th December A delightful story in the latest edition of the bee-keeping journal of Saxony-Anhalt. A brown bear was being sent from Glatz to Halle by rail. During the journey to Eger the bear managed to break down the partition which divided him from other goods in the van. He consumed two geese, several kilos of butter, and a basket of cherries. Two baskets of eggs were trodden to pieces and scattered about, as was a parcel of margarine. On arrival at Eger the railway official opened the van, and the bear made to receive him with open arms. The official fled to find the consignor, who quieted the bear by giving him – a piece of honeycomb.

Which reminds me: Jas. Hiams told me that in 1804 his great-grandfather, also called James, gave William Pitt a piece of honey-comb when he, Pitt, then Prime Minister, came along the south coast inspecting the new defences against the French. This happened at Seaford.

Friday, 13th December I can hear Fiona practising carols below with Mlle. Zabiellska.

This morning, top-dressed the asparagus beds and did as much pruning of fruit trees as the cold would allow me.

Friday, 27th December A sociable Christmas, with Sir Alan Russell, Thornley and Dolly de Groot, Dr. Jawalhir Lal Sinha and his wife, Sonakshi, and Mrs. Mountshaft, on a very cold, but sunny, Christmas Day. Mrs. Trench and Pearl both being with their families, Fiona, Dolly and Peggy Mountshaft shared the cooking in the morning: partridge pudding, home-grown vegetables, fruit pie and plum pudding. We all bade the bees Merry Christmas after lunch, and then the men washed and dried the dishes while the women walked round the village.

I gave Fiona a gold, moonstone and sapphire brooch made by John Paul Cooper, and she gave me a book – *Father and Son*, just published, but anonymously – which I shall begin shortly. Sir Alan spent the night of the 25th with us, having enjoyed too much of one of my best brandies to be able to go home.

I also received a wonderful book by post, from William Heinemann Ltd. in London. It is *Syria: The Desert and the Sown*, by Miss Gertrude Bell, who asked her publisher to send it to me in appreciation of the pleasure she has derived from reading about me in the *Strand Magazine*, while she has been away from the comforts of the *beau monde*. She has travelled throughout the Ottoman Empire, and through Syria, photographing and recording her findings. She is a diplomat, an archaeologist, a linguist fluent in Arabic, Persian, French and German, and a mountaineer. In 1902 she apparently spent forty-eight hours clinging to the rock face when a violent storm struck the Finsteraarhorn as she was climbing it.

I usually end each year in this journal with some bee notes of a general character, but this year I have none ready. I do have some preparatory notes on the swarming of bees, which I hope to include at the end of next year.

1908

Friday, 3rd January Yet another extravagant party at Bateman's on New Year's Eve. Kipling received the Nobel Prize for Literature in early December, and his house overflowed with well-wishers. I can remember the food and drink well enough – cold birds, ham, tongue, foie gras, gold patties, jellies, ices, champagne and – my particular nemesis that evening – rum punch. I clearly recall standing on a chair at one point and reciting – to applause – the villages of the East Sussex Weald in alphabetical order, namely: Alciston, Alfriston, Arlington, Berwick, Best Beech Hill, Birling Gap, Blackboys, Blackham, Bodle Street Green, Boreham Street, Broad Oak, Burlow, Buxted, Chalvington with Ripe, Chelwood Gate, Chiddingly, Coleman's Hatch, Cooper's Green, Cowbeech, Cross in Hand, Crowborough, Danehill, Duddleswell, East Dean, East Hoathly with Halland, Eridge Green, Etchingwood, Exceat, Fairwarp, Five Ash Down, Five Ashes, Fletching, Folkington, Forest Row, Framfield, Frant, Friston, Furner's Green, Golden Cross, Groombridge, Gun Hill, Hadlow Down, Hailsham, Halland, Hammerwood, Hankham, Hartfield, Heathfield, Hellingly, Heron's Ghyll, Herstmonceux, High Hurstwood, Holtye, Hooe, Hooe Common, Horam, Horney Common, Isfield, Jevington, Laughton, Litlington, Little Horsted, Little London, Lower Dicker, Lower Horsebridge, Lullington, Magham Down, Maresfield, Mark Cross, Mayfield, Maynard's Green, Milton Street, Muddles Green, Ninfield, Nutley, Old Heathfield, Pevensey, Pevensey Bay, Polegate, Poundgate, Punnet's Town, Rickney, Roser's Cross, Rotherfield, Rushlake Green, Selmeston, Stone Cross, Stunts Green, Tidebrook, Three Cups Corner, Uckfield, Upper Dicker, Upper Hart-field, Vines Cross, Wadhurst, Waldron, Warbleton, Wartling, West Dean, Westham, Whitesmith, Willingdon, Wilmington, Windmill Hill, Winton, Withyham, Wych Cross. *Chalvington with Ripe. Muddles Green. Three Cups Corner.* It was the closest to po-etry I shall ever come, but I fell off the chair at the end. The band struck up for the dancing and Kipling took me outside to where he has his own small watermill, and I doused my head with cold water.

Fiona and I danced, and then in the dawn I found myself on a sofa with Freeman Freeman-Thomas and others, expatiating on the im-portance in detective work of using the senses quickly: noting phys-ical features, particularly eyes, ears and noses, and the influence of trades on the form of the hand; the nature of the physical objects on a person, particularly dress and shoes, but including rings, watches, and walking sticks; personal mannerisms; accents and the use of dialect words and expressions; perfumes, and powders and eau de Cologne. I went on to talk of animal characteristics, particu-larly their tracks and prints, and including human shoe and boot prints, bicycle tyre prints and carriage wheel tracks, and no doubt very soon, automobile tyre tracks; effects of the weather; dried blood; the infinite variety of handwriting, inks and writing paper; the smell and detritus of different firearms and explosives; differ-ences in typewritten documents; and the varieties of pipe, cigar and cigarette ash on which I have written a small monograph – *Upon the Distinction Between the Ashes of the Various Tobaccos* – in which I distinguish 140 varieties. To my chagrin I found that Free-man-Thomas had been asleep for most of my discourse. Fiona and I came home at 6a.m. with the Mintys in their automobile.

Monday, 6th January A great joy – a black cardboard box with the label "Meccano 2" – meaning, Outfit No. 2 – and the inspiring words, "This Outfit contains everything necessary to make com-plete working models, including all tools and full instructions –

Cranes, Bridges, Towers, Aeroplanes, Wagons, Signals." My birthday present from Fiona.

As we sat smoking after my birthday lunch, Freeman Freeman-Thomas asked me about the case of Wilson, the canary trainer. He said he did not know there were such men as canary trainers. Well, Wilson *was* a canary trainer, and a cruel one. There was an extraordinary fashion in the 18th and 19th centuries to train birds to sing popular music on command. Bullfinches, linnets, canaries and other songbirds were taught popular tunes as well as songs written specially for birds. Once trained, these poor songsters were used as feathered music machines, delivering music in people's homes. Birds with the most mellow or musical tones were selectively bred to produce cocks with even better songs. Some of those with the best songs were then used to train young cocks, so that the talent could be used in two ways. Indeed, some of the birds were sold exclusively as trainers, others as songsters.

It is not difficult, I told him, to breed canaries, but to produce fine, prize-winning singing canaries is a question of expertise and dedication. Every young cock must be trained before he becomes a useful and calm singer. The training begins during the breeding in the nest and continues in the singing cage.

The case was brought to me in 1895 by a young journalist who had found on the pavement in Finchley one morning a parcel from which was issuing, to his astonishment, birdsong. He picked it up. In all his experience of country rambling, he told me, he had never before heard such distinct and emphatic bird-music, crisp, sharp, and ringing out at regular intervals, as though the tiny creature from whose throat the sounds proceeded were a mechanical toy. There were no wasteful and extravagant flourishes of melody, no whimsical jumble of notes short and notes long, with wanton twitterings between, such as a free bird delights to indulge in – here I remember his exact words – but "a shrewdly calculated and systematic performance, as though after every renewed effort, he wound himself up for the next, and was bound to deliver himself to the instant of a certain quality and quantity of music, as per contract."

This journalist unwrapped the parcel and saw a finch in a bamboo cage. He realised, to his horror, that the bird had been deliberately blinded. It is common belief among trainers that the darker the singing birds' environments, the better they sing, and as a result many breeders permanently destroy their songbirds' eyesight with red-hot needles, or wire. In the parcel's wrapping the journalist found a letter to the purchaser which gave him the names of two bird-catchers. His investigations led him to the bird market and animal shops in Sclater Street, E., and to one in particular – Wilson's. I heard described to me in detail the appalling conditions the birds were forced to endure. This was the "plague spot" to which Watson referred, and which I was able to eradicate after investigating Wilson's criminal past.

Thursday, 16th January We have had a spell of severe frost – 18F – and in consequence the bees have been confined to their hives.

Thursday, 30th January My brother, Mycroft, seven and a half years older than I, died on the 19th of this month of a seizure, and on Tuesday we buried him in the London necropolis at Brookwood in Surrey. There was a small crowd there standing in the snow, among whom I noticed William Melville, and, in *mufti,* Sir Neville Lyttleton, the veteran of Tel el-Kebir, Omdurman and Spion Kop. I said a few words at the graveside, as did the Secretary of State for the Home Department, Herbert Gladstone; the president of the Diogenes, Campbell Laird; and Mycroft's lifelong companion, Stuart Crawford. A choir of boys from Eton sang Newman's *Lead, Kindly Light.*

Thursday, 6th February Right on to the end of January we had practically no snow, yet the weather has been unsettled, with rough winds, keen frost, and a variable temperature. Bees have been much confined and had few thoroughly good flights. Never have I seen more dead thrown out.

Friday, 14th February I read in the newspaper of a forthcoming piece of government legislation – a Children's Act – which fills me with hope. It will establish separate juvenile courts, and require the registration of foster parents, which will reduce baby-farming and infanticide. It will also grant to the new "Local Authorities" powers to set up their own orphanages, and will keep children out of the workhouse, prevent them working in dangerous trades, and bar them from purchasing cigarettes and entering public houses. Dickens, thou shouldst be living at this hour! I say again, legislation such as this will do infinitely more to abate crime in our country than all its police forces and consulting detectives, no matter how determined and clever.

Tuesday, 18th February Ice-skating on the pond at Telscombe. I mentioned Dickens in the previous entry. My ice skating would have earned me a place alongside Mr. Winkle's immortal efforts in the *Pickwick Papers.*

Tuesday, 25th February Planted some border plants, and delphiniums, peonies and ranunculus.

Wednesday, 11th March Planted artichokes, asparagus and potatoes while Fiona planted hollyhocks – what a joy for the bees! – and phloxes and carnations.

Thursday, 26th March This late, almost flowerless Spring points to the need for a supply of pea-flour for a few days till pollen can be gathered.

Thursday, 2nd April April has arrived and weather conditions remain bad. Cold NW winds, interspersed with hail, sleet, rain and thunder; and what of the poor bees during such a long spell of inclement weather? A glimpse of sunshine brings them out, and then the cruel storm beats them down in hundreds, never to rise again.

Monday, 6th April An unusual weekend at Ambrose Gorham's place to celebrate his daughter's birthday. We must have been fifty in the party, at least. Yesterday afternoon all the men were made up and dressed as clowns by a team of young assistants, and then we performed to a script written by Gorham, all to amuse the young lady. We did this in a tent erected in one of his paddocks and dressed as a circus within, with a band in attendance, our womenfolk being the audience.

I have been in two minds about the inclusion of this photograph (see below) in the journal, but it was Fiona who took it, and she insists it appears. I am standing on the far right, with next to me Col. H. F. Jolly. The policeman at my feet is Gorham, with his youngest son, Bill. Minty is the terrifying clown third from the left with a white satin collar round his neck, and the absurd bald figure sitting at the front and holding a piece of wood is the Very Rev. Dr. F. H. F. Jannings, the incumbent at All Saints, Hastings, with, next to him

and kneeling, his curate, the Rev. Trevor Vardy. We were told to look solemn for the photograph, and we all did our best, except for the young man on the left of the back row, who apparently did not hear the instruction.

I hear from Arthur Harding that the Walker gang has clashed in the street with the Titanic Mob, and got the worst of it. According to Harding, the Titanics had a secret agreement with the local police, who were waiting in hiding nearby, and arrested most of the Walkers. One of them, Frank Cooper, aged fifteen, had a loaded gun on him, which is not an offence in itself, of course, but it caused all those arrested to receive a week's remand for affray.

Tuesday, 14th April In the Post Office in Polegate I was shown a parcel covered in exotic stamps and with only my name and the words "near Eastbourne, Sussex" on it. They know me there, but I had to fill in three forms before I could take it away. It was a book of poems by an Indian poet, Rabindranath Tagore by name, one of the latest to join this worldwide army of friends I appear to have as the result of Watson's storytelling. In a letter enclosed, Tagore tells me he lives on an ashram called Santiniketan, not far from Calcutta in west Bengal. And that at the age of seventeen he was sent to a school in Brighton, and lived in Medina Villas, Hove.

"Bees sip honey from flowers and hum their thanks when they leave," he writes. "The gaudy butterfly is sure that the flowers owe thanks to him." And, "Not hammer strokes, but the dance of water, sings the pebbles into perfection."

Monday, 20th April Easter Day was cold, dull and cheerless, interspersed with driving snowstorms. I am still supplying artificial pollen and feeding the bees with thick syrup, while providing a plentiful supply of water. If I can tide over the present untoward spell of bad weather, the promise all around of bee-forage ready to burst into bloom will sustain my hopes for better things ahead.

Sunday, 26th April Heavy snow for the last two days. On Saturday we walked on the beach at Cuckmere Haven with Miss Straightbarrel and Miss Adrienne Jones, all four of us wrapped up like the Esquimaux against the cold. We were the only foolhardy adventurers down there.

Monday, 27th April The Olympic Games begin today at White City in London.

Mulched fruit trees, and attacked weeds without mercy.

Thursday, 14th May The long-delayed Spring came in with May Day, and the cold weather has at last given way to summer heat. The bright yellow of the dandelion shows plentifully in the meadows, while the anemone and wild cherry are blossoming in abundance in the woods.

I have begun a monograph on the usefulness of wireless telegraphy in the detection of crime, and the telephone ditto. The Marconi apparatus has been installed on our warships for five years, and it is high time it was in our police stations. We have been able to exchange wireless messages with the United States for even longer. My imagination is captivated by the thought of a police station in, say, Glasgow, being able to tell a station in London to check certain shipping office records, and that London station then sending a wireless message to a ship in mid-Atlantic, say, that results in an arrest – all within an hour!

Liquid manure applied to asparagus beds, while Bill Frusher trained shoots of peach and nectarine trees.

Thursday, 28th May Dull weather, with some rain and a little sunshine, has been our weather report for the last fortnight, and its adverse influence has greatly retarded breeding. I am still feeding my stocks to prevent the bees from starving, so that even with a change for the better we can hardly hope to see any honey stored for another two weeks.

It is good to see that the excellent Mr. Israel Zangwill has changed his mind. For some years this distinguished writer and dramatist has been advocating the allocation of Palestine to the Jews as a homeland, claiming that Palestine holds nothing but a few *fellaheen* and lawless Bedouin. Now he has realised there are at least half a million Arabs there, settled on the land for centuries and farming it, and he has broken with the Zionist movement. I have written him a letter of congratulation and support.

Thursday, 4th June We are back from another short stay with the Armatradings in Dungeness. A mysterious place when the fog is down, and the only sound is the deep moan of the compressed air foghorn. How different is the fog here at the coast from the fog of London, with its grimy opacity. While walking on the shingle with the Armatrading boys I was able to correct their misapprehension, that the unpleasantness of London fogs is caused by the smoke that blackens them. The combustion of coal is certainly responsible for the fog's existence, but it is the sulphur of the coal, oxidised ultimately to sulphuric acid, and not the carbon, that is the active agent. And so long as coal is burnt at all, this manufacture of sulphuric acid and of fogs must continue; it is not to be got rid of by improved methods of combustion, though the character of the fogs may be materially altered for the better. The evil effects of town air on plant life and human lungs, also often attributed to preventible smoke, are in like manner due to this non-preventible sulphuric acid. The great gain in cleanliness and health that would follow the abolition of smoke from domestic fires in London cannot be overrated. But, said one of the boys, how would people then boil water in their homes? Electricity, I said, one day.

Yesterday the foreshore was full of people. A vessel called the *Loanda*, travelling from Hamburg to West Africa, collided in thick fog with a Russian steamer, the *Junona*, on Sunday, and was badly damaged. An attempt was made to save her but she went down under tow off Dungeness. The *Loanda*'s cargo was cases of gin, rum and champagne, and that is what the crowd were looking for, without success. There are rumours that the cargo also included thousands of newly minted silver coins.

It seems that all over the country this has been a disastrous season for apiculture. All my correspondents agree that the circumstances are abnormal and almost unprecedented. I myself have many dead bees, and many more weaklings. But the closing days of May were gloriously bright and sunny, so we may yet see a marvellous change.

Friday, 5th June I have been asked by a society of ladies and gentlemen in Lewes, who meet once a month to discuss literary matters, to address them in early July. I may take whatever topic I like. I think I shall choose, not stories from my casebooks, of which the literary aspect is the least interesting, but the popular sensational fiction of the last century. I have a *penchant* there for which I can give no account. I shall give the society – whose members' lives are, without a shadow of doubt, of irreproachable virtue – bigamous marriages, misdirected letters, romantic triangles, heroines in physical danger, drugs, potions and poisons, characters in disguise, strained coincidences, aristocratic villains and naive heiresses, suspense, dormant peerages, murderous baronets, ladies of title addicted to the study of toxicology, gipsies and brigand-chiefs, lunatic clergymen, men with masks and women with daggers, stolen children, withered hags, bitterly contested wills, heartless gamesters, nefarious roués, and foreign princesses causing havoc in the House of Lords.

Tuesday, 16th June This news of the discovery of vast quantities of oil 1,200 feet under the ground in Persia will, I predict, change the region's history for ever. It will also transform Europe's industrial and naval landscape, because whatever you can do with coal, you can do more easily with oil. And this is true without one even mentioning the infant automobile industry. The discoverers of the oil are British, and I imagine the effect on the stock market, once their company offers, will be colossal. I cannot see the British government standing aside, nor the American, nor the Russian, German or French governments. This is nothing less than a new Africa for exploitation, and I fear for the region.

Thursday, 11th June Home after three delightful days at the County Ground at Hove, watching Sussex play Kent. It was the divine Ranji I really went to see, but H. H. the Jam Sahib of Nawanagar disappointed us all. In the first innings he got 3, caught by Hutchings off Arthur Fielder, and in the second 29, lbw Fielder. He has put on weight.

Weed-killer applied to weeds on paths, with some trepidation about the bees.

Friday, 19th June This has been a good month for sunshine, but what of the honey harvest? If my colonies were full I could expect at least an average one, but with weakened stocks, and no swarms, I do not think I shall have more than half a crop. The new mowing machines in use in this neighbourhood have already laid the bulk of the bee-forage, but where the white clover is plentiful the plant will soon recover, I am sure, and throw up an abundant second crop of blossom which, with the vetches, will carry us on till the lime trees blossom in July.

Monday, 23rd June On Saturday to London. We were with a magnificent crowd of 500,000 in Hyde Park to demand votes for women. We were particularly supporting our friend Miss F.E.

159

Mintz, of Eastbourne, who spoke from the platform. The police were aggressive and constantly shouted abuse. On our way home we heard that, after the rally, twelve women gathered in Parliament Square and tried to deliver speeches for women's suffrage. Police officers seized several of the speakers and pushed them into a crowd of opponents who had gathered nearby. Angered by this, two WSPU members – Edith New and Mary Leigh – went to 10 Downing Street and hurled rocks at the windows of the Prime Minister's home, breaking several windows. I see that Emmeline Pankhurst has endorsed the action on behalf of the WSPU.

Later in the day we met Poppy Armatrading and her two sons, Daniel and Philip, who had also been at the demonstration. I took this photograph in Park Lane.

In the evening we were the guests of Mr. William Tilden at his home in Kensington. He held a party to celebrate the award to him of the Royal Society's Davy Medal for his work on the terpenes and on atomic heats. I am a chemist, amateur but passionate, and it was very heaven to be there for a while and to talk chemistry with him, and with Fittig, Dewar, Perkin, Moissan and many others. Sir Henry Roscoe complimented me on my researches into the acetones, which I have been communicating to him from time to time.

Friday, 27th June Excellent news. Gregson has been appointed Assistant Commissioner in the Metropolitan Police. I always knew he had the qualities to rise in the force. He has written to me, using the word "lucky" to describe his promotion. I am about to write back to congratulate him and also to tell him what I believe makes so-called "lucky" people, lucky. They are more skilled at creating or noticing chance opportunities than they realise; they listen to their intuition more than the average; they have positive expectations as to the outcome of challenges or problems that face them; and they are resilient in the face of disappointment, again, without their necessarily being conscious of it. Luck is a flower that grows naturally in such soil; real luck is much rarer than one thinks.

Wednesday, 15th July Another visit to Hove, this time with Freeman-Thomas, to watch Sussex v. Somerset. What should have been a three-day match was over in two, with a lamentable Somerset only able to get 99 and 94 in reply to Sussex's first innings of 404. Fry was the heavy hitter with 119. Ranji could only manage 33, but to the crowd's astonishment he took 3 for 22 in the lower order of the Somerset second innings.

Tuesday, 21st July Sowed spring cabbage, winter spinach, and planted winter greens.

My talk to the Lewes Reading Club went off very well. So many people turned out that we had to move into a larger room in the hotel where the Club meets. Modelling myself on Dickens, I gave dramatic readings from the novels I chose, including *The Woman in White*, *Lady Audley's Secret*, *The Shadow of Ashlydyat* and *Griffith Gaunt*, with Fiona reading the women's parts where necessary. I also included readings from police reports of shocking crimes – kidnap, bigamy, murder and the like – and attempted some theorising of my own about the appeal of such stories to the well-behaved classes. We ended with a lively discussion of the mountebank Louis de Rougemont's *Adventures*.

Tuesday, 28th July On Saturday afternoon we went with the Mintys in the automobile to Dover, where we stayed the night in an hotel, and rose very early to go to Northfall Meadow, just outside the town, to see the arrival of M. Blériot in his monoplane. Minty's friend, Charles Fontaine, a correspondent for *Le Matin*, waved a vast tricolour to guide the gallant pilot in after a half-hour flight from France. The landing was awkward, the nose of the monoplane burying itself in the grass and shattering the propeller, but M. Blériot was unhurt. Fiona gave him a kiss on the cheek.

Monday, 31st August Sun-tanned and slightly heavier than when we left, we are back from a fortnight's holiday in Deauville. We stayed at the Grand, where we met three interesting people. One was a fellow Englishman and bee-keeper, Dr. Hugo Merryweather, of Devizes. I learned from him that the nectar which bees suck up from flowers is secreted in a special stomach. At this point it is 80% water, with a residue of complex sugars, and not recognisable as honey. This substance is later processed by workers in the hive and deposited in droplets on the upper side of cell walls. Its conversion into viscous honey is completed by an evaporation process which is hastened by the warm temperature – 95F – maintained in the hive. Leaving nothing to chance, the bees control the movement of air by fanning their wings in a co-ordinated effort. The buzzing coming from a hive even at night, when there is no flying, is the sound of this forced evaporation process. The end result is thick, viscous honey with a 17% -18% moisture content. Pollen, on the other hand, is kept quite separate until it is mixed with honey in order to provide food for the brood – the next generation of bees.

Our other companions were a charming young couple, M. Étienne Balsan and Mlle. Gabrielle 'Coco' Chanel. He is a cavalry officer, she a singer from Moulins, in the Auvergne. She plans to open a shop in Deauville, where she will sell hats.

Friday, 11th September In London yesterday with Herr Doktor Sigmund Freud, the author of *Psycho-Analysis and the Establishment of Facts in Legal Proceedings*, which I read two years ago. I wrote to him at that time, congratulating him on a profoundly interesting and far-reaching paper, and he replied, saying that when he next came to London he would like to meet me. He speaks good English, but slowly and with a heavy accent. I met him at Ford's Hotel in Manchester Street, and took him to the British Museum, but he seemed more impressed with the shops in Kensington.

In the Chinese section of the Museum Freud and I were deep in conversation about our experiences with the stimulant, cocaine, when we were accosted by an extraordinarily flamboyant young man – tall, thin, red-bearded, with one large blue earring – in blue shirt with hand-painted tie, green trousers, pink coat, and wearing an immense "sombrero" hat. The young man had seen us peering at an inscription, and volunteered a translation for us. We gazed at him in wonder as he declaimed, waving his arms about and speaking with an pronounced American accent. He took himself off after giving me his card: Ezra Pound, 48 Great Titchfield Street, W.

This is the summary of Herr Freud's paper, as given on the back cover: "There is a growing recognition of the untrustworthiness of statements made by witnesses, on which many convictions are based in court cases. A new method of investigation, the aim of which is to compel the accused person himself to establish his own guilt or innocence by objective signs, is presented. The method consists of a psychological experiment and is based on psychological research. A word is called out to the subject and he replies as quickly as possible with some other word that occurs to him, his choice of this reaction not being restricted by anything. The points to be observed are the time required for the reaction and the relation between the stimulus word and the reaction word. It has become customary to speak of an ideational content which is able to influence the reaction to the stimulus word, as a complex. This influence works either by the stimulus word touching the complex directly or by the complex succeeding in making a connection with the word through intermediate links. The following reactions to the stimulus word can be observed. The content of the reaction may be unusual. The reaction time may be prolonged. There may be a mistake in reproducing the reaction. The phenomenon of perseveration may occur. The suggestion is that the legal profession adopt this experimental method in an effort to establish guilt or innocence by objective signs."

When we parted Herr Freud gave me a copy of his book, *Jokes and their Relation to the Unconscious*, in German "Der Witz und seiner Beziehung zum Umbewussten." I have yet to find a good joke in it.

I stayed on at the Watsons' in London to watch the boxing in the Olympics. I boxed a little in my youth, and have still have an interest in the sport. It was a splendid tournament at five weights, with Great Britain taking five gold, four silver and five bronze medals, though in fairness to the other nations I should point out that Great Britain entered 32 of the 42 boxing contestants. Richard Gunn, a Londoner, fought wonderfully and took gold at bantamweight at the great age of 37 years and 254 days, a record which I think will stand for many years. He was working in his father's tailoring business when he started boxing at the Surrey Commercial Docks Club. He was twenty-two then, and the following year became British amateur champion. After the final triumphal bout I went to the dressing rooms to congratulate him and we exchanged autographs.

In the evening I went to east London, at the invitation of George Bernard Shaw, to hear Prince Pyotr Kropotkin, doyen of the Anarchist Congress, speak to the Stratford Radical and Dialectical Club. Kropotkin lives at 6 Crescent Road, Bromley. I cannot quite say why, but that seems to me an amusing address, for an anarchist. Shaw was with him on the platform. I found Kropotkin's argument convincing – namely, that the dominant economic system under which we live, which he calls capitalism, creates poverty and artificial scarcity while hardening the advantages of the privileged. A cab ride through the streets of London is all one needs by way of evidence for that. He proposes an alternative – a decentralised economic system based on mutual support and voluntary co-operation. At the end of the evening I bought a copy of his *The Conquest of Bread*, which he signed for me, in hope of finding out how, precisely, he proposes we move from the one system to the other.

Monday, 12th October On Saturday to London again, with Fiona, to the *Ideal Home Exhibition* at Olympia. In a vast tent, demonstrations with live bees were given by Mr. W. Herrod, instructor in apiculture at Studley Castle Horticultural College. He was assisted by several of the lady students, including Emmeline Pankhurst's daughter, Adela, who recognised me.

Tuesday, 20th October Late potatoes lifted and stored, and tied tops of early cauliflowers.

Wednesday, 18th November Stored turnips, beetroot and artichokes.

Yesterday to Lady Henry Somerset's home for inebriate women in Duxhurst, Reigate, to see young Kitty Byron, who six years ago stabbed her appalling lover, Arthur Baker, to death outside the Post Office in Lombard Street. I was instrumental in raising the petition that got her sentence reduced to life imprisonment. She is out of Holloway now, but must stay in Lady Somerset's home till she dies.

Thursday, 26th November While walking west of here I stopped by a thatched cottage to greet a woman in her front garden attending to an old-fashioned skep of bees. I wrote down what she said: "My honey this season paid the rent of our cottage (£3), put boots on all the members of my household (£2 15s.), gave us several small luxuries (£1), and left us a considerable quantity of honey for home use, highly appreciated by all."

Friday, 4th December Unseasonably mild, open weather continues, and bees have been on the wing several days during the past week in such numbers that from outside appearances I consider the stocks are strong in population. I have been re-roofing my hives with thin sheet zinc.

As an experiment, we sowed some early vegetables – peas, beans and radishes – in a warm border.

Saturday, 26th December We celebrated Christmas yesterday with the Scrope Viners in Friston. We met there Barbara Scrope Viner's brother, Prinsep Normand, a coal contractor in Upper Norwood in south London, and his wife, Rose; Captain Minty and his wife, Joyce; Martin Joywheel, a painter in oils, and his companion, a Miss Donkin. We were royally feasted with croutes à l'Indienne, zéphires of duck, roast goose with braised red cabbage and turnips, and a blancmange à la vanille.

Joywheel is a collector of the traditional music and songs of Sussex. In the evening he treated us to a performance of several songs, including the Sussex Whistling Song, sung to the tune of *Lillibulero*. It has the coarsest language I have ever heard. Nothing from the streets of London's East End comes close. Joyce Minty left the room.

We exchanged presents at home in the evening. Fiona gave me a book, *The Wind in the Willows,* by an Edinburgh Scot called Kenneth Grahame who lives in Cookham. It was published earlier this year. I gave her something she had asked for, a cream duster coat for wearing in an automobile, to keep the dust and dirt off. It has blue embroidery on the collar, and large buttons of what she tells me are abalone, a kind of mother-of-pearl.

Sunday, 27th December I read in the *British Bee Journal* this morning that a worker bee lives only forty days or so. I had thought it much longer. It *is* longer, though, for bees born late in the Autumn, who live through the Winter in semi-hibernation. The article also contained precise details on the stages of the worker's life, though how this information was obtained it does not say.

Aged 1 – 2 days: cleans cells and keeps brood warm

3 – 5 days: feeds older larvae

6 – 11 days: feeds youngest larvae

12 – 17 days: produces wax, builds comb, carries food, undertaker duties

18 – 21 days: guards the hive entrance

22+ days: flying from the hive begins, pollinates plants, collects pollen, nectar and water

Thursday 31st December The land is now covered in snow, and the thermometer outside my kitchen window registered ten degrees of frost this morning, affording the boys a happy time of sliding on a roadside pond, the ice of which is easily strong enough to bear them. The hives are but mounds in the snow, but as my stocks are packed snugly away for Winter, my mind is at rest as far as their welfare is concerned. Fiona is sitting by the fire, mending our bee veils.The weather means we shall celebrate the arrival of the New Year with our neighbours at the Tiger in the village, only ten minutes walk away. The lane is clear enough to walk on.

I received a letter today from Schoenberg, the saddest I think I have ever read. In October his wife, Mathilde, left him for a painter called Gerstl, but she was persuaded to return to Schoenberg, and at that Gerstl set fire to his paintings and hanged himself. Then just ten days ago was the first performance of Schoenberg's new string quartet, his second, in F sharp minor. The piece is apparently so bold in its gradual dissolution of a secure tonality that the Viennese audience could not tolerate it. In the second movement someone in the audience sneezed, provoking howls of laughter that drowned out the instruments. Throughout the third and fourth movements there were cries of "Stop it!" and "We've had enough!", and the piece ended amid universal protest and brutish uproar. Poor fellow! I shall of course write him a letter of sympathy, and ask to see a copy of the score, but that will be small consolation.

With all his troubles, though, this good man has found time to go through my violin-and-viola scoring of the *Goldberg Variations*, and has written to me about it in the most complimentary and reassuring terms. He got his students to play the variations to him. He says he sees no reason why he should not include the work in a concert in Vienna, when he feels ready to return to public performance. I shall have half a dozen copies printed in anticipation.

1909

Saturday, 2nd January My head has now recovered from the celebrations in the Tiger on New Year's Eve. They continued till four, at which point Constable Veevers led us out in a long line onto the snow-covered green in a wild dance which involved singing nonsensical words at the top of our voices and throwing out our left and right legs alternately. This is not how I, or Mrs. Hudson, as she then was, behaved in London. But East Dean is not Baker Street. *Tempora mutantur, nos et mutamur in illis*, my late brother used to say.

There has been much written about why bees swarm, and the control of conditions that lead to swarming. It should be remembered that with bees and other social insects the community is the unit, rather than the individual. The workers are incapable of reproduction, and accordingly no matter how great an increase there may be in their numbers in a hive, it is but temporary, and makes no permanent difference to the perpetuation of the species. Swarming is thus the natural means of reproduction of honey bee colonies. The queen bee leaves the old colony with perhaps half the worker bees when conditions favour the creation of a new colony. Swarming is mainly a spring phenomenon, at the height of the honey flow when major nectar sources are in bloom and the weather is clement, but occasional swarms can happen throughout the producing season. Enough honey, or nearly, will already have been brought to the old hive to carry that colony through the Winter, and at this season the new swarm should be able to leave with a minimum of danger and establish itself elsewhere.

During the first year of a queen's life the colony has little incentive to swarm, unless the hive is very crowded. During her second Spring, however, she seems to feel an instinctive urge to swarm. Without the beekeeper's swarm management in the second year, the hive will cast a "prime swarm" and then may cast one to five "after swarms." The old queen will go with the prime swarm, and other swarms will be accompanied by virgin queens. All this is in the interest of the safety of the bees, of course, but the bee-keeper sees his colonies divide at the very time he wishes the them to re-main large and increase his honey crop. So, when he sees the bees begin to "hang out", as the expression is, in large clusters outside the hive, for want of room, he knows he needs to act, and will place supers in his hives to give the bees more storage space, replacing each super as it is filled up.

It is my practice to clip the wings of the queens to prevent their flying off with the swarms, but for this method to be effective I need to be constantly on hand at the critical time. The swarm issues, the air is full of bees, and I go to the hive from which they have issued and usually have little trouble in finding the queen moving about in front of the hive. It is then an easy matter to place her safely in a cage, remove the old hive and put a new one in its place. The bees – if they have clustered in a convenient spot, which is not always the case – may then be shaken into a basket and dumped in front of the new hive at once. I then release the queen and she runs inside with them.

Clipping is a decided advantage in a garden like my own, where fruit trees surround the apiary. It is no easy task to try to capture a swarm gathered in a tree, perhaps twenty feet off the ground. For those bee-keepers who do find themselves scratching their heads and gaz-ing up into the higher branches I can recommend the ingenious Swarming Ladder designed by H.Robert Strimpl, of Selzschau in Bohemia. Herr Strimpl kindly sent me a kit of one of his ladders, which I assembled with ease. At its upper end is a single pole that can be lodged securely in the fork of an upper branch, where a two-pronged ladder could not. Three downward-pointing iron prongs at the bottom of the ladder hold it securely in the earth and prevent rotation. The bee-keeper simply mounts the ladder with a swarm-catcher. Herr Strimpl recommends for this purpose a large wire-cloth cage about two feet long and 12 to 15 inches wide. This seems to me in every way superior to the tiresome machinery of Mr. A. E. Manum's Swarming Device and the preposterously complicated Hiving Pole of Mr. Miles Morton.

Even the best-laid plans for rehiving, however, may be upset by the return of enthusiastic scouts. These bees have been out searching for new premises for days in advance of the issuing of the swarm, and may have found somewhere they consider even more comfort-able than the bee-keeper's new hive, and may well persuade the new colony to abscond on the following day. I am glad to say I have not yet had this disaster befall me. William Tompsett tells me that this is the result of my hives being placed, on his recommendation, in a cool, shady spot.

In addition to the clipping of the queens, beekeepers may control swarming prior to the natural swarm time by removing frames of brood comb, making 'nucleus' colonies. (The term 'brood comb' refers to the beeswax structure of cells where the queen bee lays eggs. It is the part of the beehive where a new brood is raised by the colony. The brood comb is usually found in the lower part of the beehive, while the honey comb may surround the brood area and is found exclusively in the honey supers.) Nucleus colonies are small honey bee colonies created from larger colonies. The term "nucleus" refers both to the smaller size box and to the colony of honey bees, with their queen, within it. The nucleus box is a smaller version of a normal beehive, designed to hold fewer frames. The box is smaller because it is intended to contain a smaller number of honey bees, and a smaller space makes it easier for the bees to control the temperature and humidity of the colony, which is vital for brood rearing. When using a Langstroth hive, a nucleus is cre-ated by pulling two to five frames from an existing colony. These frames and the nurse bees clinging to them form the basis for the nucleus colony. A nucleus is extremely vulnerable, as it possesses in some cases less than a tenth of the workers in a normal colony, and its bees almost always need to be fed. Feeding allows the worker bees to remain in the nucleus, acting as nurse bees for the developing brood. The bee-keeper must be doubly watchful in the early stages as the nucleus is vulnerable to robbing, where a stronger hive steals all the nectar, honey, or syrup from a weaker hive. The bees from a robbing hive will kill any bees that defend the nucleus, and robbing can lead to starvation in days, but a well-protected nucleus can easily grow into a full-sized colony, given proper time, favourable weather, and appropriate resources.

Thursday, 7th January An unexpected birthday present from Scotland Yard – a Fortnum's hamper, no less, containing Scottish smoked salmon, venison salami, marmalade-glazed ham, a formidable Stilton, a tin of chocolate biscuits, opulently fruity marmalades, a bottle of Bollinger *Grande Année* and a superb Sancerre. And a corkscrew.

Saturday, 9th February Attended this morning a sad ceremony on Upper Street – the placing of a stone over the burial place of the favourite dog of the Fuller family. The dog, an excellent hunter of rabbits, was run over by a farm wagon. The stone reads simply: SPOT JAN 18:09.

Monday, 22nd February What a delight the past four days have been. I got my first coltsfoot yesterday, called coughwort by some of the older people. Snowdrops have been abundant for a week, daisies are plentiful, and the spear points of the crocus are coming up in impressive array. Yesterday I heard larks over the fields towards Charlton.

Planted anemones.

Monday, 8th March Last week ended in a blizzard that nearly equalled in severity the great blizzard of 1881. There were 36 inches of snow in Eastbourne, and the railway cutting near here was blocked by a drift ten feet deep. We were cut off for two days of glorious, quiet isolation, but today the snow is nearly melted, the sun is shining and the bees are humming. Is there another country in the world with such madcap weather?

Tuesday, 16th March We are baking bread for the first time, guided by Mrs. Trench's mother.

Wednesday, 24th March Bill Frusher has helped me make and erect a small dovecote as a lean-to against the south side of the house. Now to find some doves.

Tuesday, 6th April When making my first spring examination yesterday I found my stocks much more forward than I had expected; brood-rearing is advancing well, with all stocks having three or four combs with a good patch of brood, and the work is still merrily going on. I do not remember having seen the bees work harder than they have this last fortnight.

Monday, 12th April An Easter of walks over the Seven Sisters, hot cross buns, simnel cake, and a salutary experience for me. On Saturday I took an early swim in thick mist, plunging into the sea without a care and swimming strongly for a while. Then I stopped, and realised that I had no idea in which direction the land lay. There was no sound of waves breaking to guide me, and no indication of the position of the sun. If I chose the wrong direction I could find myself swimming far out to sea. Then, as I trod water and pondered my fate, the mist thinned a little and I saw the sun, which I knew was in the east at that time of day. I turned and swam north, keeping the sun at my right hand side, and before long heard the waves breaking faintly ahead. What a death that might have been, swimming aimlessly until exhaustion overcame me! I thought I heard Moriarty laugh at the thought of my joining him in that way.

Sunday, 2nd May May has come in with a shining face, though with cold easterly winds which, in exposed positions, have retarded the work of the bees. Myriad blooms of dandelions make the fields to the east of here appear as if covered with cloth of gold.

Placed clean straw round strawberry plants.

Friday, 14th May I have been to London three times in the past ten days. A new bureau is being set up as a joint initiative of the Admiralty and the War Office. The bureau's role is to ensure national security through counter-espionage, which is why I, and the head of the Special Branch of the Metropolitan Police, have been consulted. Its name is to be the Secret Service Bureau, though by the time this journal of mine is published, if indeed it ever is, I do not imagine there will be much secret about it. I anticipate its principal activity will be the identification and neutralisation of foreign agents, mainly Russian and German.

Sunday, 16th May A letter has come from Los Angeles, from a bioscope director called Mr. D. W. Griffith. He wishes to make a film of Watson's highly-coloured case history, *The Hound of the Baskervilles*, and to have me go out there to advise him. He has sent a photograph (see below) of an actress whom he wishes to have a part in the film, a "Gibson Girl" called Mary Pickford, who he says is keen to meet me. I am writing to tell him that the rights in the story belong either to Watson or to the *Strand Magazine*, and that he should give my compliments to Miss Pickford, but I shall not be going to Hollywood.

Thursday, 27th May Fiona has won £12.16s by backing *Minoru*, the King's horse, to win the Derby, on the advice of Ambrose Gorham. I am kicking myself for scoffing at the idea.

We are now in the midst of the fruit-blossom – apple, pear, and currants black, white and red. The damson and gooseberry blossoms are just over, and there promises to be plenty of fruit this year if no late frosts in May cut the young fruit.

Planted out lobelia and pyrethrum.

Saturday, 29th May Bee colonies are being laid waste throughout the south of England, I read, by "Isle of Wight disease", which was first noticed there first two years ago. Opinion has it that it is the result of either an intestinal parasite or a mite. This area is free of it at the moment, as it is, incidentally, of foul brood also.

Sunday, 13th June To the Hurlingham in Fulham with Fiona on Saturday to see a point-to-point ballooning race. We went up in the automobile with Captain Minty and Joyce, and helped Minty's son inflate his balloon with hydrogen. (As a chemist, I have the gravest doubts about the use of hydrogen, that most flammable of gases, in this way.) Fourteen balloons took off in bright sunlight in the late afternoon, a splendid sight. Apparently a German won, and received £80 prize money.

Friday, 18th June The outlook for a good honey season in this locality is fairly bright. The welcome rains of late have improved matters greatly, and the sainfoin appears to be in excellent condition as far as blossom goes. Jas. Hiams does not remember a Spring when colonies have built up as quickly as this one, due, he thinks, to the liberal amount of honey left in the hives last Autumn, the young queens, and the warm spells of weather during which honey and pollen were gathered plentifully. We have had quite a honey glut from the early blossoms, and two of my stocks have filled a section-rack. The honey seems to be of excellent quality, though not of the delicate flavour characteristic of that gathered from sainfoin. The apples are coming well into flower. Jas. Hiams, who is in a mood to impart information at this time, says that the success of sainfoin on these soils is the result of its long, tapering roots which descend into fissures of the chalk, breaking it up and rendering the ground deeper and better for cultivation. He will make a country-man of me yet.

I have been invited by Robert Baden-Powell to attend a rally of his "Boy Scouts" at Crystal Palace. He would like me to talk to them about my techniques of deriving significant information from small, often unnoticed evidences in people and locations. I think I shall not go; the very worst way of communicating my methods would be by standing on a platform and haranguing a crowd of young men in the heat. I have offered to write him a pamphlet he can have printed and distributed.

Tuesday, 15th June Two days ago, when walking over Mill Down to Wigden's Bottom, I noticed close to the bridleway a pretty cottage which had at the side some eight frame-hives. It was a beautiful evening. Wondering what race of bees he stocked, I went into the garden and found the owner amongst his roses. He told me had Carniolans, but had lost four swarms on May 24th, all issuing on a Sunday, he said, "while the church bells were ringing."

Planted out tomatoes and cucumbers, and Bill Frusher transferred celery into trenches.

Monday, 21st June Early on Sunday evening some persons tried to take the honey in an apiary in our neighbourhood. The thieves placed a sulphur match at the entrance and suffocated the bees; they then took the combs, which they put in two buckets. But the buckets, the honey and other things were found in a heap in the road outside the apiary; that was the only evidence, apart from the dead bees. No-one else was about. Why would the thieves have done this? I was consulted. Once on the spot I made an immediate diagnosis. The thieves had forgotten that at that hour on a sunny day not all the bees were in the hive. I pictured those which had been gathering honey, on returning home to find such a desolation, beginning to make a great fuss, which would have attracted the inhabitants of other hives, and in a moment the thieves would have been surrounded by a multitude of furious insects. They were compelled to take flight, and abandoned not only their spoil, but also their buckets and other things, in the road. What's clever in that?

said Jas. Hiams, who was one of those with me. I smiled, remembering Watson's same reaction at hearing the end point of some simple chain of deduction.

A story from Miss Ada Dell, an old lady in the village. Some years ago an ancient horse, nearly useless from rheumatism, was left near her house in the week. Suddenly, during breakfast time, a considerable noise attracted her attention; the bees of a nearby apiary were attacking the poor horse, which was vainly contending against them. After he was delivered from his assailants he recovered easily from the stings, and from that time, until he died two years later, he was completely cured of his rheumatism.

Thursday 24th June To London for the Royal Agricultural Society's Show, Bee and Honey Section, where Fiona and I showed in the *Open Two 1lb. Jars of Light or Medium Clear Honey* Class and in the *Open Honey Cake* Class. No luck this time, but the competition was formidable, so we are not downhearted. We met Mr. George Bernard Shaw at the Show, surrounded by adoring females. All vegetarians, presumably.

Monday, 28th June Yesterday we were blessed with a fine day for a trip by the paddle steamer *Brighton Queen* from Eastbourne pier along to Bognor Regis and back.

As we were returning home we saw a travelling fair on the move, heading west – for Brighton, I suppose. Mighty Showman's Engines were lugging the equipment and the rides – swingboats, roundabouts with their gallopers, and gigantic things wrapped in tarpaulins advertised in gold letters as "Razzle Dazzle", "Waltzer", "Speedway" and "Steam Switchback".

Saturday, 10th July Last month was one of the most disappointing Junes on record, wet and dull days predominating, with bursts of sunshine of short duration, and July, so far, is much the same. Bees have been confined to their hives for the greater part of the last four weeks. The barometer is steadily rising, though this means a contest between the farmers, who will soon be cutting their grass in earnest, and the bees, who will want to be at the white clover before it is cut down.

Tuesday, 13th July I have been filmed. And not just filmed, but filmed in colour – an astonishing thing. Mr. George Smith, who lives along the coast in Southwick, on the other side of Brighton, introduced himself to me about three months' ago. He came this morning with his equipment and two young assistants. He has made previous films with this process, which he calls Kinemacolor – *Tartans of Scottish Clans* and *Woman Draped in Patterned Handkerchiefs*. All I had to do was to come around the corner of my house, walk towards the camera and raise my hat. He will show me the result in a week or two.

Sunday, 15th August July is gone, but the month will live long in the memory of bee-men in this locality as the coldest, windiest July ever experienced. One might count on the fingers of one hand the days on which the bees gathered any honey. In early July hopes were high for some days; then the bad weather set in. But the weather improved in early August. There was a profusion of blackberry blossom half a mile distant, and from about 6 to 9 a.m. the roar of the bees overhead in their flight to and from this forage astonished the passers-by, the bees coming home with fully-distended honey-sacs.

Monday, 23rd August Planted strawberry runners in new plantations, and cut away old fruiting canes of raspberries.

Monday, 30th August Some few days' indisposition have allowed me to catch up with my scrapbook work, and complete the revisions of two entries.

179

I have mentioned earlier in this journal the Cleveland Street brothel scandal of 1886. Yesterday evening I reached the name "Arthur Newton" in my scrapbook, and was reminded of the small but significant role he played in that unsavoury affair. When Charles Swinscow, a fifteen-year-old Post Office messenger boy, was accidentally discovered by a policeman to have eighteen shillings in his pocket, the boy said he had earned it at 19 Cleveland Street, London W. He told the policeman that that address was a homosexual brothel run by a man called Hammond, and frequented by aristocrats. Later, in the police station, Swinscow named names – Lord Arthur Somerset, the Earl of Euston, and an army colonel called Jervois. Somerset was the son of the Duke of Beaufort, a major in the Horse Guards and in charge of the Prince of Wales's stables. After being interviewed by the police, Somerset consulted his solicitor, Arthur Newton. Realising Somerset was in serious trouble, Newton went discreetly to the Assistant Director of Public Prosecutions, Hamilton Cuffe, and told him that if Somerset were to be prosecuted, or even mentioned in connection with the case, he would name Edward, Duke of Clarence, the Prince of Wales's eldest son, as a regular user of the brothel. This trick worked; neither Somerset nor Euston was prosecuted, and Somerset prudently left for France. I had all this from Somerset himself, who foolishly attempted to enlist my help on his behalf. Naturally, I declined to be involved, and in any case I was retained at the time by the Prime Minister and the Foreign Secretary in the case Watson called "The Second Stain."

The boy who had started all this was arrested, along with three others. He received nine months hard labour; the others, four, and then the matter was forgotten. Two years ago, a man called Parke used the *North London Press* to name Somerset and Euston as the aristocrats not named at the boys' trial. Euston sued Parke for libel, claiming that he had indeed been to 19 Cleveland Street, but only under the misunderstanding that it was a conventional brothel. When he discovered what it really was, he said, he had left in a hurry. But Parke's counsel produced witness after witness to say that Euston had been there many times – was a "regular", in fact. The judge fulminated against these witnesses, calling them "loathsome objects", and the jury found Parke guilty. He was sentenced to one year's hard labour.

What travesties of justice! Newton was involved in the defence of Oscar Wilde in 1895, and of Dr. Crippen last year, with a distinct lack of success in both cases.

Saturday, 18th September Wasps are very numerous this season, and the persistent way they dodge about the entrances of the hives, seeking admission for a taste of the honey if they can manage to pass the guards, is amusing, especially when the bees bundle them out and off the alighting-board without ceremony.

Planted phloxes and violets.

Sunday, 26th September This from the latest edition of *Le Rucher Belge: Bulletin de la Société d'Apiculture du Bassin de la Meuse*: 'Lors de l'exposition apicole de Breslau, 20 août, un apiphile national découvrit, dans ses pérégrinations à travers la ville, un arbre couvert d'une masse de fleurs blanches et sur lesquelles toutes les abeilles de la ville semblaient s'être données rendez-vous. Après enquête, il réussit à savoir qu'il avait affaire au Sophora japonica, arbre à croissance très rapide, prospérant dans toutes les terres, donnant un bois excellent et, ce qui le rend précieux au point de vue apicole, produisant une profusion de fleurs depuis le commencement d'août jusqu'en septembre. Il est mellifère à l'égard de l'acacia et permet encore une abondante récolte de pollen. M Kramer, dans le *Schw.Bztg.*, compte cet arbre parmi les plus mellifères, méritant surtout l'attention des apiculteurs dont les abeilles n'ont pas de miel tardif.' But for me, certainly, and for most of my readers, it is too late to plant one! It would be hopeless to expect flowers on a Sophora japonica less than forty years old.

Thursday, 30th September To London, at the suggestion of M. Rodin, to see the works – two paintings and twelve woodcuts – of a Russian, Wassily Kandinsky, in the Royal Albert Hall. While we were admiring them there was a stir in the crowd and the King appeared. I had been pointed out to him, and he was kind enough to spend some minutes in conversation with Fiona and myself. He was accompanied by Sir Lawrence Alma Tadema, whose complexion, after viewing the paintings, was an alarming mixture of green and purple. I attempted some words in defence of this 'abstract' art, but the King snorted, and moved on.

Saturday, 2nd October As an instance of the value of bees, I may mention that last Spring, after five weeks of cold weather, we had one very warm day. The pear blossom had been kept back by the cold, but during the morning about six trees burst into full bloom in the most astonishing manner. The bees simply stormed them, and by nightfall almost every petal had dropped off. The next day was wet, and the cold again returned. As a result, the trees are full of fruit, due, to my mind, only to the rapid fertilisation of the blossoms by the bees. Other than these, there are few pears this year in our immediate district.

Monday, 4th October Yesterday evening I heard, while walking back by Cornish Farm from Frost Hill, three sharp barks of a dog fox, the plaintive whistle of the golden plover and the musical cry of the curlew, then later the screaming and hooting of owls.

Tonight I reviewed my scrapbook entry on Charles Peace, dead these thirty years – hanged at Armley Prison, in Leeds, in February, 1879 – but still one of the strangest criminal minds in all my records. These are the barest details. He was always a skilled burglar. At the age of 22 he wounded a policeman, and killed another twenty four years later, to escape arrest, all that time living and working in Sheffield as a picture framer and seller of musical instruments. Then he shot and killed a man called Dyson, whose wife he had fallen in love with. He fled to London with a price on his head, and set up home as Mr. John Ward in a respectable villa in Peckham, along with his wife, Katherine, who knew nothing of his murders, and his mistress, Susan Thompson. By night he burgled houses in the Blackheath area, mainly, and it was there he was eventually caught. When he saw the game was up he confessed cheerfully to all his crimes, including the murder of the policeman, so that the man who had been wrongfully convicted of that crime, and was languishing in prison on a life sentence, could be freed.

Friday, 8th October The honey season would have been a reasonable one had not the limes failed in July, owing to the dull and cool weather that prevailed just then.

With Bill Frusher opened and cleared gutters, gulleys and drains.

Sunday, 17th October On Saturday up to London in my very best attire, with Kipling, to a lunch reception at the *Strand Magazine*. I sat next to a very shy young man, Mr. Pelham Wodehouse, a writer for the musical comedy, who is beginning to have stories published in the magazine.

Monday, 18th October The weather continues cold and unsettled, and is likely to remain so for a time. It is an unusual sight to see corn still standing in the fields so late in the year; in fact, I saw the reaping machine at work today, cutting oats. Apparently it is many years since the ingathering has been so late. A neighbouring farmer told me he had to re-make his ricks, as the rain had penetrated them before they were finished, with the result that in the centre the corn was rotting. Freeman-Thomas, who has a vast library, tells me he thinks this has been one of the coldest Summers since 1659.

Thursday, 21st October A disappointing audience last night – eleven – for my lantern-slide lecture, "Bees in Relation to Flowers and Fruit", in the Eastbourne Technical Institute and Free Library, a green-and-white-tiled palace of learning for the common people.

183

But the weather was appalling, and the common people stayed at home.

I have received a letter on interesting note paper. It is from a Mr. B. Timms, who says he needs my help in the matter of a robbery. He would like to meet me next week in London, in Brown's Hotel, at some time convenient to myself. I must be in London next Thursday on other business, and, life here being so quiet, I have, in the modern slang, a "yen" for an investigation, so I will meet him.

Friday, 29th October Yesterday, after a stroll in delightful autumn sunshine in Green Park, I went to Brown's at 4pm. A young man of aloof manner and extraordinarily smart appearance introduced himself to me in the lobby and conducted me to a suite upstairs. I was surprised to see that the room which we entered was divided by an enamelled screen. The young man said simply, 'Your visitor, Mr. Timms,' and left us.

'Good afternoon, Mr. Holmes,' came the voice of a man of about my own age, or perhaps a little older, from behind the screen. 'Please sit down. Thank you for troubling to come to London to see me.'

'It is my pleasure,' I said. 'I was in London anyway, and this is one of my favourite hotels. May I ask why you do not wish to be seen?'

'My appearance is thought odd,' he said, 'and is something of an embarrassment to me. Please indulge me in this anonymity.'

'As Your Majesty wishes,' I replied, with as much gravity as I could muster.

'"Your Majesty!"' he said. 'Whatever do you mean? My name is Timms. Have you taken leave of your senses? What on earth makes you think I am the King?'

'Sir,' I said, 'it was not difficult, when I received your letter, to note that the paper was of exceptionally fine quality. I then held it up to the light, which revealed at once the lion and the unicorn of the royal watermark. I am familiar with more than six hundred watermarks. Next, the young man who brought me to you has "royal aide-de-camp" written all over him, if I may use that expression. His exquisite dress, his even more exquisite manners, the very slightest bow he could not quite restrain himself from making, as he left you – all these traits of the courtier confirmed my original suspicion as to your identity. Then, there is in this room the unmistakeable aroma of a cigar recently smoked. My nose tells me it is a *Hoyo de Monterrey*, from Vuelta Abajo in Cuba, widely known to be Your Majesty's favoured brand. I would venture further that it is a grand corona, and possibly a *Geniales*, but that is by the by. Lastly, I met Your Majesty exactly one month ago, at an exhibition of paintings by the Russian, Wassily Kandinsky, in the Royal Albert Hall. Today, on entering this room, I recognised your voice at once.'

'Dammit, Mr. Holmes, you make it sound so simple, a child could do it!' he said. 'What was that case of yours where the jewel was found inside a goose at Christmas?'

'*The Blue Carbuncle*,' I said.

'That's right. Knowing what you could deduce there, just from looking at – what was it? – an old hat, I should have known better than to imagine I could fool you. Bring your chair round here, please.'

He shook hands with me, and I noticed that he looked rather ill – thinner than recent pictures in the newspapers, with a grey pallor under his beard, and an ugly scar next to his nose. His generous suit seemed a little big for him.

'To business,' he said. 'It's theft, Mr. Holmes, as I told you in my letter. Five pieces of jewellery that belonged to my mother, the late Queen. Two tiaras, two brooches and a necklace. They disappeared a fortnight ago, from our private apartments at the Palace. My wife and I were somewhat lax there, I suppose. She had worn one of the tiaras and the necklace the evening before, but had laid out all five pieces before deciding which to wear, and they were still on her dressing table the following day, rather than in the safe where they should have been. They seem to have gone at some time during that morning after the ball.'

'In cases like this,' I said, 'it is almost always the rule that a servant is the culprit.'

'I am sure you are right,' he said. 'I myself questioned four of them who would have had the readiest access, but they are all of long service and of the most impeccable honesty, I am convinced.'

'You have informed the police?'

'No. They were wedding presents to my mother. One of the tiaras was a gift from Czar Nicolas I, and the necklace has as its centre-piece a yellow diamond given to the Queen by the Maharajah of Travancore. To lose such items, with such connections, is a grave embarrassment. Once the police know, I imagine it would only be a matter of time before the story was out and in the newspapers. And besides, the thief has written to me.'

'The thief has written to you?' I said, in astonishment.

'This is the letter,' he said, reaching to an inside pocket of his jacket. 'He mentions you.'

Do not worry, it read. *You will have them back, but you must pay. Sherlock Holmes will be your go-between, and no one else. We trust him to play fair. The meeting will be in Hyde Park, by your father's monument. Announce agreement in* The Times. *No tricks!*

'I take it, Sir,' I said, 'that you would like me to procure these items for you, if I can?'

'You are correct. I suppose, as you are one that bestrides, so to speak, two worlds – polite society, and the criminal underworld – that the rogue is quite right to nominate you. You are a rare bird in that respect, Mr. Holmes – a very rare bird.'

'And what is an acceptable price?'

'Whatever the price, I shall meet his terms. But if, with your un-paralleled powers, you were able to recover the jewels without pay-ing him, you could expect the most generous tokens of my admira-tion, and gratitude, as well as any fee you cared to name.'

Both envelope and letter were cleverly written – the cheapest sta-tionery, block capitals in pencil, no trace of a regionality in the grammar or vocabulary. And, no mis-spellings – a feature which, I thought, might be worthy of note.

'How did the letter come to you?' I asked.

'In the normal way, through the post. My secretary handed it to me unopened because it has "Strictly Private" on the envelope, as you see.'

He began to cough violently, and for some moments we did not speak.

'He asks for notice as to the time and date of the meeting to be placed in the Personal Column of *The Times*,' I said, when he recovered. 'Is that acceptable?'

'Of course,' he said. 'Shall I provide agents hidden your place of meeting, to get hold of him?'

'Sir, that would be most unwise. He will undoubtedly have made provision for that eventuality, and immediately after his arrest you would find the world informed of your loss – the very thing you wish to avoid. You are in deep difficulty.'

'Which is why I rely on you, Mr. Holmes!' he said, with such vehemence that another fit of coughing came on. I left him shortly afterwards, took a cab to the offices of *The Times* to place my advertisement, then went on to London Bridge Station.

Tuesday, 2nd November At eleven yesterday morning I went up the steps to Prince Albert's memorial. There was thick fog; I could see barely five paces in front of me. I waited some twenty minutes, examining the grandiose statues at the memorial's base – bare-breasted Asia on her elephant, Africa as Cleopatra on a camel, the spirit of America on a bison, and Europe crowned and riding on an ox amid her symbols of arts and power. Then suddenly a small body materialised out of the fog, and in front of me stood a street urchin in a ragged green coat.

'Name of 'omes?' he said.

'Holmes,' I said.

'First name, Sherlock?'

I nodded.

'For you,' he said, and handed me an envelope.

I grabbed him by the shoulder.

'Ten shillings for you if you tell me who gave you this,' I said.

'Keep your money,' he said calmly, and in one beautifully prac-
tised movement swung his arm up inside my grasp and then brought
it down again hard against my wrist, breaking my hold. Then he was
away down the steps and gone. Prince Albert seemed to gaze down
on me sardonically through the fog, amused by my feeble efforts to
help his eldest son.

The envelope was blank, and the note written in the same style as
the first. *You were too slow,* it said. *They are sold. They are in Vienna now. If you want them, go there. Stay at the Imperial and you
will be contacted.* I was not overly surprised by this message. There
are dealers in stolen goods of the highest value – they are called
fences, in thieves' slang – who buy from burglars, greedy servants,
corrupt officials, and the like, and make highly profitable business
that way. I have dossiers on many of them – Nicholas Purkis, du
Maurier *père et fils,* Leeuwenburgh, Gollancz, von Mensdorf-
Pouilly – but none of them in Vienna. I must tell the King I cannot
go there for him. He must find another messenger.

Friday, 5th November I was admitted early this morning to the
King's apartments. His Queen Consort broke down in tears as he
spoke to me of the impact the news of his loss would have on his
reputation, and I relented. I shall go to Vienna for him. I have just
explained all this to Fiona, who is the soul of understanding and

concern. She wishes to come with me, but I would not hear of it. I have sent a telegram to Watson.

Saturday, 6th November Yesterday evening we hosted a fireworks party for the children of the village. There was a fine bonfire, a guy that I could swear looked like Lloyd George in a top hat, bread and sausages, cocoa and biscuits, and a selection of Pain's best fireworks – rockets, Catherine wheels, fire barrels, radium cascades, Roman candles, rainbows, snowflakes, Vesuvius cones, jumping jacks, and bangers of all descriptions, including a monstrous thing called *Aerial Artillery* that had me fearful for the bees as they slept. No reply from Watson.

Sunday, 7th November A gratifying scene yesterday afternoon in the auction rooms of Burstow and Hewett, in Battle. I was there to bid for a handsome Black Forest cuckoo clock for Jas. Hiams and his wife, for Christmas. I had seen on the back the label, *Johann Baptist Beha und Söhne*, which told me it was from the finest manufacturer in the world, and not from a mass production house like Junghans or Philipp Haas. I won the item – there seemed to be only one other bidder, and he lost interest when my determination to have it became clear – and then the auctioneer asked me my name. When I said it, there was a round of applause in the room! – and much hand-shaking afterwards.

Watson has replied. He will come with me! I knew I could rely on him. God bless my old friend for his loyalty and bravery. He is important enough in his profession these days to be able to find a locum at a moment's notice. There may be little or no detective work to do here, but it will be wonderful to be on the trail with him again.

Monday, 8th November With Bill Frusher pruned apple, plum and pear trees, and gooseberry and currant bushes, in the rain.

Monday, 15th November This weekend in the Soup Kitchen I heard about a bakery in New York City's Greenwich Village called the Fleischmann Model Viennese Bakery. For years they have had a policy of distributing unsold baked goods to the poor at the end of each business day. There must be bakers in Eastbourne and Hastings who can follow Greenwich Village's example. I shall investigate.

After a fairly good start to November, cold, wet and boisterous weather has returned. In my little apiary two of the covers were blown off and the wrappings saturated with rain, but none of the hives was overturned, so that no serious harm resulted.

These two brothers on the left were on the wall outside the Soup Kitchen when I arrived on Sunday morning. They had walked four miles in bare feet to get there. The young dandy on the right – all alone in the world, he told me, but still able to smile for my camera – had been waiting by the back door since dawn. No parental care, no school, and no protection, in the mightiest and richest country the world has ever seen. And prison awaiting them with their first theft. I wonder if Ulyanov, if he ever comes to power in Russia, will so change their system of society that boys like these, and girls, will get a better chance in life?

Monday, 29th November I doubt very much that these jottings of mine will ever be published, and so the astonishing events of the last two weeks will remain unknown to the world at large, at least until after I am gone. I have refused point-blank Watson's offer to record them, though he pleaded with me. Fiona is writing these words at my dictation as I lie with half my hair burned away, ointment on the burns on my face, and both my hands bandaged. I have not taken the laudanum this evening – I will take it later – so as to have my mind clear enough to recount the tale.

And what a tale it is. Watson and I crossed on the Newhaven to Dieppe ferry on the morning of the 18th, and left Paris on the modestly-named "*Orient Express de la Compagnie Internationale des Wagons-Lits*" at 6.30 that evening. We dined on oysters and turbot in green sauce in the neighbourhood of Rheims, and drank champagne before retiring as we approached Strasbourg. We were in rare spirits. At noon the next day we strolled about the Munich Hauptbahnhof while our train was stopped there, and I introduced Watson to *bratwurst* and mustard with *sauerkraut*, which were not greatly to his taste. We talked so much during the afternoon – of marriage, London, modern medicine, bees, politics, the dangers and triumphs we had shared over the years – that the noble landscapes *en route* to Salzburg went by virtually unnoticed. We were still talking as we pulled into the Vienna Hauptbahnhof towards ten in the evening.

There was serious business to do the next morning. Our arrival had been observed. My breakfast arrived with a note under the coffee pot. It said simply, *Two million Kronen in gold coin. Wait.* I had with me an open Banker's Draft from the King, and when I showed it at the Rothschild Bank the money was brought up without demur, in four wooden boxes tied with leather straps. Two guards from the bank came with us back to the hotel, where the boxes went carried into the strongroom under the manager's astonished gaze. And then, as instructed, we waited.

At about three, as we sat over whisky in the hotel's saloon bar, a cabman came into the room and asked for me. I went upstairs for my coat as the boxes were brought up to a side entrance and loaded into the cab. I got in with them. I had already given Watson his instructions: to have his Army revolver with him, just for safety's sake, to follow us in his own cab until we stopped, then to wait twenty minutes, and at the end of that time, if I had not emerged, to come into wherever I was, and find me.

It was just beginning to get dark as we trotted along the Burgring towards the Josefstadt quarter, and there were flurries of snow in the air. The horse kept up a spanking pace and soon we were out in the countryside. The air was biting cold. I was thankful for the blanket the cabbie threw me from the trunk, and pleased that the snow would help to keep Watson's cab unobserved. After half an hour's driving we stopped at the lodge gate of what seemed to be a large country estate. Two men, muffled to the ears in ulsters and with their caps pulled down low, opened the gate and then got up with the cabman. We went along a winding drive, at first through meadows, as far I could see, and then through trees, and then the crunch of hooves and wheels on gravel told me we were arrived. It was what in France would be called a *château*, tall and turreted, of the eighteenth century as far as I could make out, with stone steps and a pair of ornate balustrades, all now evenly coated with snow, leading up to the front door. But, I noted, not a light to be seen in any window.

I felt perfectly composed as I stood on the gravel while the men carried the boxes up into the house. Fences, as a breed, detest and fear violence, and I anticipated none. I patted the cab-horse on the neck and assured him he would not have a long wait in the cold, and then a servant with an oil lamp appeared at the front door and beckoned to me. I went in.

The house was cold and dark. I followed him up a formal staircase, along a corridor hung with paintings whose subjects I could not make out, then up more stairs and onto a landing facing a pair of handsome double doors. The servant opened them and, without a word, showed me into a cavernous room with no furniture except a side table on which stood a single oil lamp. He withdrew, closing the door behind him. The windows were shuttered, but through the panes above I could see the snow swirling against the glass. As I approached the table my shadow, grotesquely enlarged, darkened the walls and spread across the ceiling. I folded my arms, and waited.

I had not stood there above five minutes when a door in the darkness at the far end of the room opened, and I saw a figure come in, pushing someone in a wheelchair towards me. Sitting in the wheelchair, her hands folded in her lap, was a woman I recognised. It was Lucy, Moriarty's widow, whom I had last seen in the cellar of her house in Sussex. I bowed my head in mortification that she had now tricked me not once, but twice.

'Mrs. Moriarty,' I said, summoning up my last reserves of composure, 'again, it seems, I must congratulate you. I thought I was coming here to collect some jewellery.'

'My mother cannot speak,' said the veiled figure behind the wheelchair. I knew the voice at once as that of Emma, the daughter. 'She suffered a seizure four years ago, some days after the death of her son, an event which you will clearly remember. She cannot speak, or stir.'

'May I ask,' I said, 'how this ambush was arranged?'

194

'It was I that planned everything. How the jewels came into my possession, you need not know. We still have our networks, as you do. I knew that once I had the jewels, they were the bait that could draw you to us. I calculated that you could not refuse a request from the King himself, and on such a sensitive matter. It was clever of me to recommend you to him, was it not? And because my mother could not come to you, I had to entice you here.'

'You are indeed your father's daughter,' I said.

'The King has paid for the jewels, and he will get them,' she said. 'They are of no importance.'

She paused.

'Mr. Holmes, I could have taken your life more than once in the time that has passed since we last met. Why did I not? Because my father would not have wanted that. That must be my mother's act, not mine. Tonight, at last, it is she who will make the final move in the game you thought was won.'

With great deliberation, she took from her dress a pair of tongs and moved with them to the table where the oil lamp was standing, then picked up a piece of cloth with the tongs and held it over the flame of the lamp until it was alight. Holding the flaming rag aloft, she began to push her mother towards me.

'Now!' she shouted.

As the cry rang out, a concealed door in the paneling of the room by me was flung open, and a man appeared, holding in both hands a bucket filled with liquid. The strong smell of petroleum came to my nostrils. Before I could move, he threw the contents of the bucket into my face. I was drenched, and fell backwards heavily, choking and wiping my eyes. When I opened them, the wheelchair was at my side. As I began to struggle to my feet the daughter, in one swift movement, took hold of her mother's hand, wrapped the fingers round the tongs, lifted the lifeless arm and shook the flaming rag down on to me. The mother looked straight ahead, utterly expressionless. At once I was a mass of flames, on my clothes, my hair, my hands and face. I rolled over and over, vainly trying to beat them out, and then – the distant sound of a gunshot, a splintering crash, and a familiar voice!

'Holmes! Dear God, man, keep still!'

Watson was smothering me with his overcoat. I was aware only of the pain in my face and hands, but that for just a few seconds, for I then passed out. Watson tells me he put me over his shoulder and went down the stairs with his left hand holding me secure and his right hand brandishing his revolver. If this escapade were ever published in *The Strand Magazine*, that would make a good illustration for the story, would it not? His cab took us back to Vienna, and he got me to a clinic where they treated my burns with characteristic Austrian efficiency. They told me when I left a week later that my hands, with the correct exercises, would eventually recover their function, and that, though the skin on my face is now permanently discoloured, my aquiline looks would survive. Fiona is smiling at that as she writes. I will have a bare patch at the back of my head for the rest of my life, but – what do I care? There are such things as wigs, and, thanks to Watson, I *have* the rest of my life.

I have written to the King a brief account of what has happened; now all we can do is wait to see if the jewels appear. I suppose we will never know what has become of Lucy and Emma Moriarty. When the Austrian police, alerted by Watson, went to the big house the next day, they found no one in the house. The boxes of gold Kronen were gone. In respect of my involvement with the house of Moriarty I have now drawn three times on my account at the Bank of Good Fortune. For fear of being overdrawn next time I shall never again go alone to an assignment, no matter how strong the temptation.

Saturday, 4th December I am at last able to write for myself, although the handwriting is appalling. Yesterday a gale grounded the *SS Eastfield*, 2150 tons, between the Belle Tout lighthouse and Beachy Head. It was en route from Hull to Barry, in Wales. No-one was hurt – the crew walked off the ship and along to Birling Gap at low tide. It is expected that in a day or two she will be pulled off at high tide. Today I tottered down there with Fiona to see her, and we had tea with the coastguard in his cottage. He told us that Birling Gap was raided in Saxon times by Danish and other pirates – "fiercer than wolves", he said. There is still some commercial fishing from here, mainly sprats, and mackerel in Summer, and there are many lobster and crab pots still to be seen. For mackerel, the fishermen wait till they see what they call "a school" – a splashing in the water out to sea, sometimes half a mile long – and then they take the boats out.

Thursday, 9th December Results from my bees this year are not much worse than my neighbours'. I have had 80lbs. of extracted honey, good in point of flavour and density, and mainly clear in colour.

My hands much improved. The bandages are off my right hand.

Thursday, 16th December This is the third of three days of fascinating microscope work on my lichen collection. Only those who are in the habit of using a microscope can have any idea of the marvels hidden beyond the reach of ordinary vision, and of the wonderful beauty and perfect forms of nature's minutest structures. I was privileged some months ago to receive from a leading optical goods manufacturer of Japan a prototype of their first compound monocular microscope, fashioned in brass with a black-enamelled horseshoe base. It came as a token of the appreciation of help I was able to render the government of Japan, in October of this year, in the matter of the shocking assassination of Japan's Resident-General of Korea, Prince Ito Hirobumi. This microscope, a marvel of modern technology, has a revolving nosepiece holding three objectives that are interchangeable with standard screw threads. The flat, square stage is fixed to the limb with a plate attachment. A substage Zeiss Abbe condenser, composed of two lenses, one plano-convex and the other bi-convex, collects light from a double mirror mounted on a tailpiece. The limb is connected to the base with a pivot mechanism and contains the rackwork for the fine and coarse focus adjustments. There was a time when this magnificent instrument would have been used in my examination of evidence in criminal matters; now it will be for bee-disease diagnosis, pollen analysis, and the study of honey bee anatomy. Examining the skeleton of a bee with this instrument is like walking over the skeleton of a whale; with the help of two set-screws the object under the lens can be moved back and forth, or from right to left, gradually and with the greatest ease. Under this same microscope minute pollen grains look like a pile of oranges of large size, having the same rough appearance.

My old microscope, which I shall now donate to our local school, was a gift from Dr. Ernst Abbe of the Zeiss Foundation in Jena, in 1898. It came with a pine box, courtesy of the German Embassy in London, which included ninety professionally-prepared microscope slides of insect studies. That microscope was of invaluable help to me in solving the Isadora Persano case, because I could see in unprecedented detail the worm that was at the centre of that affair. It was also crucial in the business of Bishop Cosmo Lang's curate, about eight years ago. Lang was Bishop of Stepney at the time, and I was able to identify the most minute particles of zinc and copper in the seams of the trousers of one of his curates, who was running a coining factory in Poplar. My Benn and Franks monocular compound microscope, my first, and which served me well in the '80's and '90's, could not have done that.

Thursday, 23rd December To Eastbourne last night with twelve friends to see *Aladdin*, followed by supper. An excellent show, with a Flying Palace, a market-place in old Pekin, an Enchanted Cave, and a Jewel Ballet of sixteen dancers. What more could you ask for 2/6? Mr. Ted Young, who is the Widow Twanky, is an East Dean man, so we applauded him loud and long. One amusing piece of business – towards the end of the first act a policeman appeared at the front of the stalls, protesting that the piece had too many wise-cracks about the local force. He was so convincing that at first we all took him for the real thing. He then fell into the orchestra pit, to huge laughter, climbed up onto the stage and confronted Aladdin, the diminutive Miss Marie Elsie, and then confirmed he was part of the entertainment by going down on one knee and singing a duet with her in a rich baritone.

Tuesday, 27th December On Christmas Day after church we entertained the Mintys, the Gorhams, the Misses Sturdy and the visiting minister, the Very Rev. F.H.F. Jannings, come over from Hastings. In the evening we were joined by Jas. Hiams and his wife, and Mr. and Mrs. Leeke and their children, for a cold buffet, charades – for which I am developing a considerable talent – and the singing of carols. I caught sight of myself in the mirror over the fireplace, carolling away with the best of them, wig firmly in place, and thought – well, it is a thought I have had many times, and to repeat it may be tiresome, but it is nonetheless a true one – as the Psalmist says, *The lines are fallen unto me in pleasant places.* On Boxing Day morning I helped the children with their Zag Zaw puzzle – *In the Hayfields*, 750 pieces.

Fiona gave me a collection of novellas, *Erma Bifronte*, by Signor Luigi Pirandello, recently published in Milan. I gave her a painted Noah's Ark made in about 1820, I should say, in the Erdgebirge region of Germany, carved in the country manner. There are 120

199

figures, including Mr. and Mrs. Noah, their three sons and the sons' wives, the raven, the dove, and even the olive branch. I saw it in a shop in London at the end of September, bought it secretly and had it sent down by train.

Wednesday, 29th December The old year is fast drawing to a close. The trees are bare, and the branches stand out darkly against the winter sky. I am in my overcoat in my laboratory, exiled from the downstairs of the house by amateur dramatics rehearsals. Panto-mimes, Variety shows, concerts, charades – I enjoy them all, but for some reason I cannot abide amateur dramatics. The play is *The Shortest Day,* by Hugh Darnby, to be put on in February.

Two quiet midwinter days with bright sunshine have provided yet more excellent opportunities for microscope work. In studying the musculature of the bee, I have found that each group of divergent fibres starts from a tendon. The axial cavity of this tendon and the hypodermis that covers it show that its method of formation is by invagination of the tegument. Each fibre must be considered as a cell with several nuclei. The sarcolemma of the fibre represents the cellular membrane. The tube formed by the sarcolemma is extended by a semi-fluid hyaline and homogeneous mass, into which the longitudinal and radiating filaments enter. The latter connect together the longitudinal filaments, and the semi-fluid substance serves as nutriment to the fibres bathed by it. Under nervous excitement I have seen the longitudinal filaments contract locally and in sympathy with each other, while the radiating ones are extremely elastic and support the others, in order to transport the nervous movement, and to bring them back to their original position after contraction.

My other recent study, of the mandible, indicates that it is an arrangement of very varied organs. Besides the usual filiform organs, I have detected a number of umbellate forms which do not appear to have been hitherto noticed. I am in the process of describing and illustrating these. I think they may be sensory organs of chemical perception, i.e. organs of smell, differing entirely from those in the antennae. It could be that they come into use in connection with the elaboration of wax, and the collection of pollen and propolis.

And, finally, the year's weather having been so poor, I append the official weather statistics for 1909, like a damning school report:

Rainfall: 36.90 in (rain fell on 188 days) [Heaviest fall: 1.67 ins, October 28]

Sunshine: 1,852 hours [Brightest days: May 19 and 30, both 14.3 hours]

Sunless days: 54

Maximum temperature: 82F on August 12 and 13

Minimum temperature: 11F on March 3

1910

Thursday, 1st January Last night Fiona was the "star turn", as they say in the Follies, at the New Year's Eve festivities in Eastbourne General. The hospital found a piano from somewhere, and she played for an hour without stopping while the junior staff, and even some of the patients, danced. I was her page-turner. Some of the gaudy sheet music is in front of me now:

Slow Drag Two Step: Hard Times

Ragtime: Easy Money

Ragtime: Come Clean

Nervy George

Ragged Edges

New Orleans Ragtime: 'Taint No Use

Louisiana Ragtime: Dynamite Rag

Can You Beat It? Rag

Friday, 7th January We ate so well on my birthday yesterday that I have adopted a New Year's Resolution – a régime of Swedish Drill exercises, on the Ling System, in the garden, at seven every morning. I am by temperament a late riser, so this will require all my willpower, which, fortunately, is still considerable.

Fiona gave me a wonderful birthday gift – in a fitted wooden case, a three-inch brass refractor telescope, with tripod and other accessories. Now I have instruments to make the invisible, visible, and the unimaginably distant, near. But I wish I could, like Prospero, command the clouds to clear.

The last of the hand-bandages is off. My hands look rather like the claws of a chicken, but they work.

Thursday, 13th January With the delightfully mild, dry and pleasant weather during the first week of 1910 the bees had a romping time of it, and the marked evidences of their faeces on and around every hive showed that the relief to them was very timely. Now, starting fresh, with all their stores rearranged, and the winter cluster ready to contract again with the threatened approach of cold, they should be in the very best form for standing a siege from whatever part of Winter still lies before us. We are not out of the wood yet. I have examined my quilts and coverings, and removed one that was damp and mouldy, and I have placed a three-pound cake of candy on the top of the frames in one hive where I suspect more stores may be needed.

My wife tells me she now avoids looking out of the French windows early in the morning. My *Free Leaps, Compound Leaps, Dancing Steps* and *Forward Leaps* in the garden are unnerving.

Saturday, 15th January The earliest flowers of the year are the most precious to me. The Christmas rose is already in blossom, with its innumerable golden stamens. Winter aconite is a hardy and cheerful flower that breaks through the frozen ground at this time. I have noticed two great scarlet blossoms on the Japanese quince in the garden of a friend, Col. H. F. Jolly. The sweet-scented coltsfoot, or winter heliotrope, is growing wild on the Downs, and there is a field of charlock near here, over Birling way, so bright and fragrant that I am sure no early-venturing bee could resist it on a sunny day.

Monday, 17th January I received today a parcel from Greece, containing two jars of honey tightly packed in straw. This late-arriving Christmas present is from Kyrillos Makarios, a Patriarch of the Coptic church, whose brother I was able to help some ten years ago. One jar contains thyme honey from Mount Hymettus, with an aroma that makes me dream at once of bays of limpid blue water, fringed with pines. The other is honey from wild roses in Zante, the most southerly of the Ionian islands. It is highly aromatic and, I am told, without equal in the world.

Makarios also sent in the parcel an old Greek coin showing a bee, which I may set on the title page of this journal, if and when it is published as a book. The goddess Artemis and her priestesses were often represented as bees, he tells me. As the symbol of Athena at Athens was an owl, so the symbol of Artemis, or Diana, at Ephesus was a bee.

Tuesday, 25th January Last night I found in Samuel Pepys's Diary, for 1665, a reference to a glass hive, thus: "After dinner to Mr. Evelyn's, he being abroad, we walked in his garden, and a lovely noble ground he hath indeed, and among other rarities a hive of bees so as being hived in glass you may see the bees making their honey and combs mighty pleasantly."

Friday, 28th January A magnificent spectacle last night in the western sky, which was utterly clear and sparkling with frost – the comet. I set up the telescope and we both gazed long at it, and with some difficulty I got Jas. Hiams and his wife to put on their coats and leave their cosy cottage to admire the magnificent visitor with us. It was between Aquarius and Pegasus, the head like a small white nebula of the 3rd magnitude, with a bright, gently curving tail.

Sunday, 30th January Today has been a most beautiful day. The bees were flying freely about noon and I took the opportunity to take a look under the quilts of each stock. I was surprised at what I saw. My bees were getting very close to the end of their stores of food, even though they were well fed up for the Winter at the beginning of October, some with syrup and some with natural stores.

The Winter here has been so mild that clearly they have been able to leave their clusters and take flight from time to time, with the result that there has been greater consumption of stores than usual. I shall give each stock a cake of candy to be on the safe side.

Thursday, 3rd February A letter has come from the Palace. I am to present myself to the King in three weeks' time. My wife is invited to accompany me.

There is hardly a greater pleasure for the bee-keeper than to see his bees in February going down into the golden crocus cups, whose transparent sides show him the gatherers at work as though through a lighted window. A bed of crocus flowers is a veritable pollen mine to the bees on a bright day in early Spring. Already open, though only in ones and twos, are the blossoms of my hepatica, some a sky blue, others pink, and, of course, the snowdrops are out. I have identified three main groups from their differing leaves – some have simple strap-like leaves, some have folded or rolled-up leaves, and some have leaves with a pleated edge.

Wednesday, 9th February Pancakes for all our neighbours for Shrove Tuesday yesterday, with our own extracted honey.

Saturday, 26th February We put on our glad rags, as they say in America, and went together to the Palace yesterday afternoon. We were received in private by the King and Queen and had tea with them for half an hour. We heard that the missing jewels were returned a month ago, by an urchin in a green coat who brought a parcel to the back gate of the Palace, put it down in front of the sentry there, and ran off. Surely it is the same boy I met by the Albert Memorial – obviously a highly-trusted member of the Moriarty Irregulars. I was made to recount, in graphic detail, the story of my adventure in Vienna, and then the Queen rang a bell and an equerry came in with a gold chain, which the King put round my neck. It is the Royal Victorian Chain, a thing normally reserved for the royals to award to each other, but apparently bestowable on some few commoners who have done outstanding service for the monarch. I have it on the table before me as I write. Its badge is a gold, white-enamelled Maltese Cross with a central medallion bearing Queen Victoria's cypher and her crown, all studded with tiny diamonds. It is a very great honour, I am told, but it confers no style or title, and must be returned when I die. I shall wear it to Kipling's next party, and cause a sensation. The King also gave me a ruby tie-pin to give to Watson in appreciation of his saving my life, and asked me to convey his compliments to the good doctor.

Saturday, 5th March The month of March arrived in quite lamb-like fashion, a whole week of fine weather coming as a welcome change. My bees have fairly revelled in the blossoms.

Sowed grass seed where the lawn is thin.

One of the most detestable of crimes is blackmail. I have been reviewing the entry for Charles Augustus Howell in my scrapbook. He is now twenty years dead. His *modus operandi* was to begin a correspondence with someone he intended to blackmail, pretending to share their sexual, financial or other weaknesses. He hoped thereby to obtain compromising letters in return, and when successful he would paste the letters into an album. Soon the victim would be told that Howell needed money, and had pawned the album. If the victim could not send Howell the money to redeem the pledge, the pawnbroker would no doubt soon sell the album on the open market. The victim's money was always forthcoming. Howell tormented dozens, including Dante Gabriel Rossetti and Algernon Swinburne, in this fashion. I was not surprised to hear, while I was engaged in the entertaining case which Watson called *The Red-Headed League*, that Howell had been found dead in the gutter outside a public house in Chelsea with his throat cut, and a half-sovereign between his teeth, which is what they do with blackmailers.

Sunday, 6th March I read something odd this evening after dinner – Charles Darwin, writing of himself as a baby in a cradle: "The windows of my mother's room were open in consequence of the unusual warmth of the weather. For the same reason, probably, a neighbouring beehive swarmed, and the new colony, pitching on the window-sill, was making its way into the room when the horrified nurse shut down the sash. If that well-meaning woman had only abstained from her ill-timed interference the swarm might have settled on my lips, and I should have been endowed with that mellifluous eloquence which, in this country, leads far more surely than worth or honest work to the highest places. But the opportunity was lost and I have been obliged to content myself through life with saying what I mean in the plainest language."

Tuesday, 15th March I see in the *Times* that the Chinese military commander Zhao Erfeng has marched into Lhasa at the head of 2,000 troops and assumed control of the government. Once again, the 13th Dalai Lama has had to flee Lhasa. This time he has gone to India, intending to take a boat to Pekin to make another attempt to make peace with the Qing court.

Wednesday, 23rd March The bees had a good flight today, working on willow, arabis, daphne and coltsfoot. Thrice blessed is he who has a tract of damp soil near his apiary growing willows bearing rich bloom in early Spring! No other flower or plant at this season is worth naming in the same breath with the various families of salix. The bees revel on the catkins, a delightful and contented throng crowds around them, and a stream of bees steadily march to and from the hives during every daylight hour.

Tempted by the fine weather we cycled for three hours over Lullington Heath.

Saturday, 26th March Today I heard – and then saw – my first humblebee this year, a perfect spring day with hailstorms coming off the Channel between bursts of bright sunshine.

Monday, 28th March (Easter Monday) In church yesterday morning the sermon was from Luke 24, verses 41 – 43: *And while they yet believed not for joy, He (the risen Christ) said unto them, Have ye here any meat? And they gave him a piece of broiled fish, and of an honeycomb. And he took it, and did eat before them.* A remarkably solid resurrected body, that could eat, and presumably digest, food shortly before being carried up into Heaven. How wonderfully specific and down-to-earth the gospel writers are, to say precisely, *a piece of broiled fish, and an honeycomb.* I wonder who kept the bees that made that honeycomb. Or perhaps they were wild bees.

After church to Alfriston with the Warwrinkas and their children, for lunch at the Star, and then we watched the old men of the village playing marbles in the street, something they have done by tradition every Easter for centuries. I firmly believe that the deep-lying English understanding and tolerance of the most bizarre practices is one of our greatest strengths as a nation, and points to a resilience that may stand us in good stead in time of war.

We saw several cottages where the local people were stitching and sewing cheap clothes, assembling wooden toys and cheap jewellery, making up bouquets of cut flowers, and pickling vegetables. We were told employment for the men is so bad that families are falling back on this piecework to keep themselves fed.

Monday, 4th April March went out with a cold wind, somewhat tempered by sunshine, but not, however, warm enough to counter the effects of a wild north-easter. The continued fine weather has provided a fine seed-time for the farmers, and a warm rain would now be welcomed, especially by bee-keepers. We are not blessed just now with many flowers. I have been giving the bees artificial pollen, and shall continue to do so for a few more days till the dandelions come into bloom.

Saturday, 16th April I am glad to report that my bees are in splendid condition, and seem strong. All five stocks came through the Winter well, and have built comb and stored honey. One stock I found so crowded on the ten frames that I gave them a super of sections and they were up working an hour afterwards, so I should get sections there from the fruit bloom. After a fortnight of glorious weather the bees are working frantically on the plum blossom and berries.

Monday, 18th April Played the violin for two hours at dawn, on the beach below the cliffs at Birling Gap.

Tuesday, 26th April Yesterday evening to Bodiam to watch the bell ringers in the performance, if that is the right word, of a Plain Bob Minor.

Thursday, 28th April We have had a succession of sunless days, interspersed with cold storms of rain and hail, which have kept the bees in their hives, but every hour the sun has deigned to shine there has been an exodus to the woods in quest of the much-needed pollen and honey. I have my supers ready to take advantage of any honey-flow that may follow May Day if the weather improves.

Saturday, 30th April With great excitement, half the village, including Fiona, Pearl Thomas and me, went down this afternoon to see the wreck of the *St. Francisco*, a Spanish ship driven on shore at Birling this morning in fog. There were no lives lost.

Sunday, 8th May The King died on Friday night. He habitually smoked twenty cigarettes and twelve cigars a day, and of late has suffered increasingly from bronchitis. Apparently, he suffered a momentary loss of consciousness during the state visit to Berlin in

February, 1909. In March this year he collapsed in Biarritz. He remained there to convalesce until 27th April, when he returned to Buckingham Palace, still suffering from severe bronchitis. The following day he suffered several heart attacks, but refused to go to bed, saying, "No, I shall not give in; I shall go on; I shall work to the end." Between moments of faintness, the Prince of Wales told him that his horse, *Witch of the Air,* had won at Kempton Park that afternoon. The King replied, "Yes, I have heard of it. I am very glad": his final words. At 11:30 p.m. on the 6th May he lost consciousness for the last time and was put to bed. He died 15 minutes later. He is to be buried at Windsor on 20th May.

Thursday, 12th May I have no improvement in the weather to report. It continues much the same day after day – dull and stormy, with very little sunshine – and of the bees that venture forth for the needful food to sustain the growing population of the hives, many never return. The approaches to my little apiary have been strewn with laden bees, their pollen-baskets packed full. In seasons of sunshine and shower the warmth of the sun revives the unfortunates the showers have beaten down, but this season we see so little sunshine that there is no chance of the bees recovering the use of their wings, and they perish by the hundreds. My wife picked up seventy bees on our garden path this afternoon. She put them in a cardboard box, with a piece of glass over the top, which she placed near the fire. Most of them soon revived and were enabled to reach the hives. She also puts food in a shallow milk-pan when we get half an hour's sun, and in the feeders if it is wet.

Thursday, 19th May Our hopes run high for a good honey season. This month is certainly a great improvement on April, but far from being an ideal May. Swarms have been scarce and not very large, or so I hear from those located in the woods and sheltered valleys in which fruit trees abound and the swarms come off early.

Yesterday to Wilmington for tea with the de Groots, and saw the astounding yew tree in the churchyard there – 1,000 years old, they say, with its two colossal trunks chained and propped up. I went into the church and found myself in the middle of a choir practice, so beat a retreat.

Monday, 23rd May On Saturday to the delightful Palladian church of St. Mary the Virgin at Glynde, for the wedding of Sir Arthur Dewhurst to Miss Silvia Fattorini, the daughter of the Italian ambassador. Sir Arthur has read all of Watson's case histories and can recall the details far better than I. When we adjourned to Glynde Place for lunch there was a professional photographer present, of course, but I was allowed to step forward with the Box Brownie and take the above picture of the wedding party.

Saturday, 11th June Today I received two interesting letters. The first was a letter of thanks from Konstantin Mereschkowski, a Russian botanist who lives near Kazan, and who studies lichens. In March I sent him forty different lichens from the stone walls in this area, each tagged with notes on its orientation and proximity to other lichens. He has a lichen herbarium containing over 2,000 specimens. I came across his fascinating work, *The Nature and Origins of Chromatophores in the Plant Kingdom*, in early 1906, and wrote to him at that time suggesting he apply the term "chromato-phore" also to the coloured, membrane-associated vesicles found in some forms of photo-synthetic bacteria. The linkages between lichens, algae, crustaceans and cephalopods are pregnant with pos-sibility.

The second letter is an invitation to visit the Shirakawa Honey Bees Rearing Garden in Zaida, in the Mitoyo District of the Kagawa Pre-fecture, in Japan. Our impressions of Japan in the late wars with Russia have been of an aggressive, over-bearing, brutal military power. How good to know that in a garden in Kagawa Prefecture kindred spirits of mine are peacefully rearing honey bees and seek-ing to improve their strains! Sadly, I do not think we can make the journey to see them.

Monday, 13th June Yesterday I spent most of the day in the garden, close to the hives. At about 3pm I saw on the platform in front of one of my hives a queen walking about and trying to get onto the alighting-board, which is some six inches from the platform. The queen was alone, but now and then a worker came and touched her, or walked round her. She was unable to fly, or at least, made no attempt to do so, although I could see nothing wrong with her wings; but she seemed a little lame in one of her back legs. I raised her on a piece of stick and put her gently on the alighting-board. She promptly entered the hive, and though I watched carefully for a couple of hours, and looked again at intervals till evening, I saw nothing more of her. Jas. Hiams told me, when I told him about this, that the queen's inability to fly had prevented a swarm leaving my hive.

Close attention to fruit trees to prevent over-cropping.

Wednesday, 15th June Yesterday Miss Nightingale, our village schoolmistress, was taking a swarm of bees to a neighbouring par-ish. The hive was placed in her donkey cart, but a few of the bees escaped and promptly signalled their liberty by stinging the donkey.

213

Naturally the donkey kicked, and soon a wheel of the cart got onto the foot-pavement, and over went the hive. The schoolmistress's predicament was such that Mr. Birtwistle, the landlord of the Tiger, went to her aid, but the bees turned their attention to him and chased him down the street. He and Miss Nightingale were stung on the head, neck and face. Subsequently the swarm attacked another man, who in trying to get away climbed over a fence and ran down a garden, where he knelt down and rolled his head in the long grass for protection, before climbing up into a tree. Others were also stung, including a passing cyclist. We bee-keepers are not popular in the village at the moment.

Friday, 17th June Wonderful scenes in the Post Office in Polegate today, as old people – those, that is, over the age of seventy, and of good character – arrived to collect their "pensions" for the first time. All of them were trying to offer produce in exchange for the pension, unable to believe the government would give them money without requiring something in return. The place was awash with fresh fruit and vegetables of every kind, pullets, dried fish, bags of eggs, preserved fruit in heavy glass jars, pies, cheeses in brown paper, jams, chutneys, pickles and many other Sussex delicacies, as well as embroidery and woven baskets. The smiling staff resolutely refused all the gifts. I was there only to buy stamps, but at the cost of a shilling came away laden with raspberries, strawberries, two jars of thick-cut marmalade and a dozen goose eggs, sold to me by women from Wilmington.

I am now sitting outside the Post Office, my booty in my lap, writing this and reflecting, not for the first time, on what has happened to me in the last few years. The confirmed bachelor is now a married man. The solitary now has troops of friends. The driven man, a user of powerful stimulants, living "on his nerves", as they say, at the summit of his profession, is now relaxed enough to sit on this bench in the sun and talk with his neighbours. The powerful spider at the centre of London's colossal web, now content with a tiny web in a Sussex garden. At first I did miss the excitement, the thrill of the chase, the elucidation of a difficult case, but that was only for a while.

Friday 1st July We are back from London after a four-day stay there with the Watsons. The highlight of the trip was the Wimbledon Ladies' Finals. Dorothea Lambert Chambers beat Dora Boothby 6-2, 6-2, after not losing a single game in any of the preliminary rounds.

Sunday, 31st July In my overly bucolic journal entry of 17th June, I spoke too soon. For the last six weeks my attention has been taken up entirely by this affair of Dr. Crippen and Ethel Le Neve which has filled our newpapers lately. I was consulted after the first inspection of the house in Hilldrop Crescent, by Inspector Dew of the Yard, proved fruitless, and such was the pressure put on me, I agreed to look into the case. Once I had spoken to neighbours, and to Mrs. Crippen's friend, Kate Roberts, nothing more than basic attention to detail was required, and I insisted on three further and ever more thorough searches, the last of which brought to light the poor woman's remains in the cellar. Only one feature of this case has any forensic interest whatever, and that is the use of the wireless telegraph in the apprehension of Crippen and Le Neve. Captain Kendall of the *SS Montrose* knew of the murder and its suspects by wireless telegram, and by the same means he cabled his suspicions of the passengers Crippen and Le Neve, who were travelling in disguise. This allowed Dew to take a faster boat and arrest Crippen as the Montrose entered the St. Lawrence River, in Canada. Dew's dressing up as a pilot to effect the arrest was pure histrionics, and quite unnecessary – almost as ludicrous as Le Neve's trying to pass as a boy. Crippen will be hanged, of course; I hope Le Neve will be acquitted, as she was not materially connected with the murder. Guglielmo Marconi and Jozef Murgas, fathers of the wireless telegraph, I salute you! The future belongs to you! Your name is Nemesis.

The men digging out the cellar in 39 Hilldrop Crescent asked me to take their picture with the Box Brownie. Here it is, also showing Dew, on the far right, and a constable of the Holloway station.

Monday, 8th August I have not had time till now to record the murder of Sir William Curzon Wyllie, aide-de-campe to the Secretary of State for India, who was shot dead on the night of Sunday, 3rd July, in South Kensington. The assassin was Madan Lal Dhingra, aged 26, a student of Mechanical Engineering at University College, London, and a campaigner for Indian freedom from British rule.

Dhingra was sentenced to death at the Old Bailey on 23rd July. He is to be hanged in Pentonville on 17th August. He made a remarkably brave and logical statement from the dock, which I reproduce here from the *Times*.

"I do not want to say anything in defence of myself, but simply to prove the justice of my deed. As for myself, no English law court

216

has any authority to arrest and detain me in prison, or pass sentence of death on me. That is the reason I did not have any counsel to defend me.

"I maintain that if it is patriotic in an Englishman to fight against the Germans, if they were to occupy this country, it is much more justifiable and patriotic in my case to fight against the English. I hold the English people responsible for the murder of 80 millions of Indian people in the last fifty years, and they are also responsible for taking away £100,000,000 every year from India to this country. I also hold them responsible for the hanging and deportation of my patriotic countrymen, who did just the same as the English people here are advising their countrymen to do. Just as the Germans have no right to occupy this country, so the English people have no right to occupy India, and it is perfectly justifiable on our part to kill the Englishman who is polluting our sacred land.

"I am surprised at the terrible hypocrisy, the farce, and the mockery of the English people. They pose as the champions of oppressed humanity – the peoples of the Congo, and the people of Russia – when there are terrible oppression and horrible atrocities committed in India; for example, the killing of two million people every year, and the outraging of our women. In case this country is occupied by Germans, and the Englishman, not bearing to see the Germans walking with the insolence of conquerors in the streets of London, goes and kills one or two Germans, and that Englishman is held as a patriot by the people of this country, then certainly I am prepared to work for the emancipation of my motherland. I make this statement, not because I wish to plead for mercy or anything of that kind. I wish that English people should sentence me to death, for in that case the vengeance of my countrymen will be all the more keen. I put forward this statement to show the justice of my cause to the outside world, and especially to our sympathisers in America and Germany.

217

"I have told you over and over again that I do not acknowledge the authority of the Court. You can do whatever you like. I do not mind at all. You can pass sentence of death on me. I do not care. You white people are all-powerful now, but, remember – we shall have our turn in the time to come."

Tuesday, 9th August It is three weeks since we had a good bee-day. This is the most dismal weather for the middle of Summer, with only three half-days, including this afternoon, on which the hum of the bees has been heard. Last week was a trying one of wet and wind, violent showers and brief, bright intervals. There are more clover and bee-flowers than I have ever seen, with the limes just coming into bloom, but the bees cannot get out to bring in their harvest without risk of being drowned, or blown away.

Yesterday we visited Holmleigh, a large country house over towards Kent with bees in the herb garden, about a quarter of an acre. The paths are edged with camomile a foot wide, lines and lines of thyme (common and lemon), several sages, with marjoram, hyssop and tarragon, and a thick holly hedge on two sides to shut off the east and north winds – truly Elysian Fields for bees.

Wednesday, 10th August Yesterday was the Eastbourne BeeKeeping Association Show. I modestly append these results:

Class III, Single Bottle of Light-Coloured Honey: 1st Prize, Master Uriah Dowsett; 2nd Prize, Mr. Sherlock Holmes; 3rd Prize, Miss Elsie Gravil; Highly Commended, the Rev. A. Frizzell.

Class VII, Honey Products: 1st Prize, Mrs. Jas. Hiams and Mrs. Sherlock Holmes (honey cake, honey liqueur, honey lemonade, honey biscuits, honey buns, honey salve, and honey fondants); 2nd Prize, Mr. Ernst Strochneider (honey furniture polish); 3rd Prize,

Miss Edith Woggett (honey table jelly); Highly Commended, Sir John Smallwood (mead, and honey vinegar).

The keeping of bees is so entrancing, the problems of a scientific nature in connection with them so engaging, and the bee-keepers themselves so numerous and varied, that even if there were no honey crop, or honey prizes, no-one who has ever really caught the spirit of the hive would willingly give up bee-keeping.

Thursday, 25th August During the last few days we have had a little more sunshine and warmth, and the bees have been busy on what forage is left. This year I have a field of mustard within a furlong of my apiary, which I hope will help with the accumulation of winter stores.

Potted early-flowering bulbs.

Monday, 29th August To Great Dixter for lunch yesterday with the new owners, the Lloyds, and their architect, Edwin Lutyens. Much conversation about their plans for the gardens and the possibility of keeping bees in them. On the way home the sky was one mass of mares' tails.

Tuesday, 30th August A late summer visit to Hastings this afternoon with the Mintys and their nephew, Hugh, to see the new American Bowling Alley on the pier. Our ancient game of skittles transformed – by American commercial genius – from a rural pastime for the old to an urban pastime for the young. We tried our best, but were all very bad at it.

The Hastings authorities have allowed mixed bathing on one part of beach for five years now, but today the cool weather seemed to have to have driven all the mixed bathers away. We walked along to the Turkish Baths at White Rock. Three hot rooms, cooling rooms and a cold plunge pool, and "marble seats and floors throughout", no less, all for 3s.6d. Fiona and I thought about going in, but then saw that Tuesday is a Women Only day.

We lunched on East parade, and on the promenade had excellent Italian ice creams from the famous cart of Signor Giovanni Montericco, who kindly allowed me to take this photograph of him, with young Hugh.

Sunday, 4th September We came upon a fairy ring while walking on the Downs yesterday afternoon. The process of formation is an interesting one. The mycelium of a fungus growing in the ground absorbs nutrients by secretion of enzymes from the tips of the hyphae, or threads, that make up the mycelium. The mycelium moves outward from the centre, and when the nutrients, such as nitrogen, in the centre are exhausted, the plants in the centre wither and die, leaving a living outer ring. I noted the presence of Calocybe gambosa, whose country name is St. George's mushroom, from the belief that it first appears on April 23rd each year. It is not on any account to be confused with the similar but highly poisonous Inocybe erubescans, or Deadly Fibrecap, which contains muscarine, a non-selective agonist of the muscarinic acetylcholine receptor. My monograph on mushrooms as poisons deals with the latter, as well as the related Deadly Webcap and Fool's Webcap, which are also rusty brown to orange. The Fool's Webcap kills, but shows no symptoms for twenty days after ingestion, which was my crucial observation in the Tarleton murders case thirty years ago, before I met Watson.

Sunday, 11th September This morning there was an episode of "robbing" in one of Jas. Hiams's hives, the first I have seen. There was a most unusual agitation of bees about the entrance. Bees could be seen running towards their enemies and quickly dragging them away. Others were fighting, and stinging each other. Bees apparently seldom rob each other when forage is plentiful, but when honey is scarce in the fields, any exposure of sweets to bees will induce robbing, and once accustomed to plunder they will try to enter every hive in an apiary. Jas. Hiams went to get a piece of wood which he placed in front of the entrance to the hive. It had a hole in the lower side just big enough to allow the entrance or exit of a single bee at a time. "To give the defenders more of a chance," he said. He told me robbing was most prevalent in Autumn, when he begins to remove surplus honey.

Monday, 26th September Gregson at the Yard has been in touch, asking for my help. Thomas Anderson, the music hall performer, was found dead in an empty apartment in London on 10th July. I declined the case. It has no features of professional interest.

Tuesday, 1st November Thanks to the unseasonably beautiful weather we have enjoyed during the past eight weeks, the face of the farmer wears a prosperous and contented look. The unfortunate apiarist, on the contrary, has had little to rejoice over, for the fine weather which was the salvation of the farmer came too late for the bees. Previously, the country around here was clothed with the blossoms of white clover, sainfoin, and charlock, which would have yielded an abundant supply of nectar had the weather been favourable and allowed the bees to take advantage of it. When the sunshine came at last, the blossoms were gone.

Friday, 11th November To London yesterday for the première of the *Concerto for Violin*, in B minor, of Sir Edward Elgar. Watson and Rekha joined us in a box, with Isabella, the widow of August Jaeger. Fritz Kreisler was the soloist, with the London Symphony Orchestra conducted by the composer. Afterwards we had supper with Sir Edward and Herr Kreisler, and a few others. I was able, from the lowly position of the amateur, to congratulate Herr Kreisler on his sublime bowing, fingering and passage-work.

Saturday, 19th November A bad business yesterday in London. With Princess Sophia Duleep Singh, the Duchess of Bedford, and Charlotte Despard, we went to support Emmeline Pankhurst and 300 members of her WSPU movement in their meeting in Caxton Hall and afterwards in the demonstration outside the Houses of Parliament. Herbert Asquith has refused to give the Conciliation Bill any more parliamentary time in the current session, and the women who would have got the vote if it had been passed are very angry. The police behaved in an undisciplined manner, brutally assaulting and manhandling women and arresting more than a hundred of them. At one point I felt compelled to intervene. In the best Association Football style, I shoulder-charged a constable who was kicking a woman lying on the ground. Without my knowing it, Fiona used the Box Brownie to take a photograph of my reckless behaviour. If the police wish to charge me on this confession they have my address and the photographic evidence of my wrongdoing.

222

Among those arrested were Miss B. L. Barwell, daughter of Major-General Barwell and Countess Elsie Leiningen; Miss Helen Craggs, daughter of Sir John Craggs; Mrs. Marshall, of Theydon Bois, daughter of Canon Jacques and niece of Captain Baldwin, the African explorer, and Mrs. Massey, daughter of Lady Knyvett.

Asquith's car was damaged beyond repair, which may cause him to ponder the strength of feeling of the women ranged against him.

I add a second photograph for two reasons. Firstly, because it shows the nobility of the Suffragette leaders, and particularly Mrs. Pankhurst, as they are led away from the demonstration by Inspector Michael Maran of the Westminster force to face inevitable imprisonment. Secondly, because it shows on the extreme right the gallant John Watson, M.D. He volunteered to go to the police station and do what he could to prevent any brutal behaviour there against the women.

Wednesday, 23rd November Two women have died from injuries received in the demonstration last Saturday. Dead, for having the temerity to wish to share in the democratic process! Words fail me.

Friday, 2nd December Wrapped exposed water pipes for the first time, on being told by Bill Frusher that very heavy frost is imminent.

Thursday, 8th December Bee-keeping has joined the twentieth century. Pictures of bee life can be seen, I read, on the bioscope at the Palace Theatre, London, nightly for two weeks, and at the Electric Theatres in Birmingham, Tunstall, Kirkgate, Leeds, Halifax and Manchester.

Monday, 19th December An urgent summons has come from Scotland Yard. I leave in a few minutes' time. It must be in connection with this wretched business in Houndsditch – a bungled robbery at a jeweller's shop, with three constables dead and one badly wounded. Three of the gang are arrested, one is dead, and the rest have disappeared. It is impossible for me to end this year's entries as I usually do, with bee-keeping notes. Once again, the game is afoot!

1911

Wednesday, 4th January Home last night, having missed spending both Christmas and New Year with my wife. But I am forgiven. She had Christmas with the Scrope Viners and New Year's Eve helping with the festivities at Eastbourne General.

It took me two weeks to track down the rest of the Houndsditch gang. They were not common thieves, but Latvian anarchists, children of the revolution in Riga in 1905. Why they attempted the robbery we shall never know. With almost all my contacts in the East End long ago gone cold, I had to resort to paying informers. I sat in saloons, drinking dens and dockside bars, but this time as a Belorussian Jew, an old prison comrade of Peter Piatkow, the leader of the gang. My story was that I had made some money from a robbery of my own in Limehouse, and was willing to dispense it generously for information as to how I could reach my friend. I spent Christmas Day in the Tidal Basin Tavern by the Royal Victoria Dock, my New Year's Eve in the Eagle and Child, Forest Gate, and I got the crucial lead on New Year's Day in the Crown and Dolphin, Shadwell, where, incidentally, they have the skull of John Williams, the Ratcliffe Highway murderer, on display behind the bar. I could not get the information to Inspector Still of the Yard till the afternoon of the 2nd, but in a drizzly dawn on the 3rd we were in Sidney Street, Stepney, in a silent cordon of 200 police. At first light they opened fire on the building I had identified, and the gang returned the fire. Soon afterwards we were joined by a detachment of Scots Guards, and Maxim guns were brought up. I was standing with a group of police in a stable entrance by the Ladies' Christian Temperance Union tea-house when I was pushed aside by a figure I recognised – Winston Churchill. He congratulated me on my work and asked me to hold his umbrella for him. He had been following events closely and, being the man he is, could not resist being in at the finish. This photograph of the two of us, with police and soldiers, appeared in the newspapers yesterday. Mr. Churchill's head is highlighted, and I am standing immediately behind him. Inspector Still can be seen crouching down at the far left.

THE HOME SECRETARY AT SIDNEY STREET

Towards noon, smoke and flames were seen coming from the building, but the door did not open. Churchill refused the fire brigade permission to enter the building, and when they did enter some time later they found two scorched bodies, but neither of them was Peter Piatkow. Somehow he had slipped away. Part of

228

the top of the building fell and killed a fireman, Charles Pearson, bringing the total of dead in this case to seven.

Churchill insisted I take tea with him in Whitehall when it was all over. I amused him by asking him what a Secretary of State at the Home Office does, exactly, and he amused me by reciting a list of his duties. I asked for it to be written down for me, and he has promised to send it.

Friday, 6th January Only the quietest recognition of our wedding anniversary and my birthday, as I feel unusually tired. Fiona and Mrs. Trench made a buffet lunch, and we invited only young Pearl, Bill Frusher, and our closest neighbours the Hiams, the Tompsetts, the Gorringes and the Misses Dandy, to celebrate with us. Quiet or not, there were eight empty champagne bottles left afterwards, and both Eunice and Ida Dandy are total abstainers. I gave Fiona a painting by the American painter Mary Cassatt, of lilacs in a vase. Fiona overwhelmed me with her lovingness, and her cleverness. My gift from her is the small study of eight apples painted in oils by M. Cézanne which I so admired on our trip to Paris in 1907. Since then she has been secretly negotiating to buy it for me, and this is the result.

An interesting letter has come, with a belated Christmas card, from Harry Houdini, actually a Hungarian called Erik Weisz, who is the world's leading escapologist. He likes Watson's tales, he says, and wants me to challenge him, publicly, by naming some predicament I believe he cannot escape from. If I do so, he intends to confound me by escaping from it, bringing, he says, good publicity for both of us. I am not to name handcuffs, chains, straightjackets, nailed packing crates, sealed boxes underwater, riveted boilers or canvas mailbags, because he has already escaped from all these many times. I wish him well, but I shall not challenge him.

This what Churchill has written to me: "I have control of the Metropolitan Police, Prisons, and Reformatory and Industrial Schools.

I supervise the inspection of factories, of coal mines, and of certified inebriate reformatories under the Inebriates Acts. I exercise certain powers under the Burial Acts, Lunacy Acts, the Employment of Children and Cruelty to Children Acts, the Explosive Acts, and the Workmen's Compensation Acts. I issue orders for the protection of certain wild birds. I grant licences for the practice of anatomy, and licences under the Cruelty to Animals Acts, and certificates for the naturalisation of aliens. I nominate the chief officers for the Channel Islands and the Isle of Man, and oversee Addresses to the Sovereign. I am also the medium of exercise of the Prerogative of Mercy in England and Wales."

Wednesday, 11th January My December efforts have caught up with me. I briefly lost consciousness this morning while dressing, and am now confined to bed by the combined forces of my wife, Mrs. Trench and Dr. Wardleworth. My breathing is also a little laboured. It will pass. Jas. Hiams came up to see me an hour ago and smuggled in a half bottle of whisky and two glasses. We drank to commemorate the centenary of the birth of Johann Dzierzon, born January 11th, 1811, in upper Silesia. If, dear reader, you have forgotten who Dzierzon was, please see my entry for 5th January 1905.

Thursday, 19th January This has been the most lovely of days. Because of the mild, open weather I see from my bedroom window hundreds of wide-open cups of the Christmas rose, showers of blossom on our winter jessamine, and stray flowers of snapdragon, polyanthus and pinks. On Saturday, Fiona tells me, the market in Eastbourne was gay with narcissi and daffodils from the Scilly Isles, and the odour of jonquils is borne to me as I write.

Thursday, 26th January The New Year has opened well for the bees. We have had fine, sunny days during which they have taken wing in large numbers. Jas. Hiams reports that they have eaten the large cakes of candy given in mid-December, and he must replenish their supply.

Thursday, 2nd February An old saying where I grew up was that if the sun shines before noon on Candlemas Day, Winter is not half over. (It was also thought unlucky by some older folk to bring snowdrops into a house before that day.) This Candlemas Day, today, the weather has been dull throughout; as to the extent of winter weather still to come, I express no opinion.

Wednesday, 15th February The weather for the past fortnight has been open and mild for February, and on several days the bees have been very busy at the watering places, and some have been carrying in pollen from natural sources. I am now completely restored to health, and, thank God, Dr. Wardleworth has disappeared from my domestic scene.

Fiona and I have brought an innovation to the Soup Kitchen. It is an American recipe called Rumford's Soup, or Economy Soup – equal parts of pearl barley or barley meal and dried peas, vegetables, four parts potato, salt according to need, and sour beer, slowly boiled until thick. One generous serving, with bread, provides a high degree of nutrition that is low in fat, high in protein and contains both simple and complex carbohydrates. It is considerably cheaper to make than what we have been providing hitherto, and more nutritious. I wish I had known about it in Baker Street days!

Friday, 31st March The bitterly cold weather of late March has kept things quiet in the apiary, outwardly at least. The month has sustained its character for varied weather, rough and cold days predominating. There have been but few days on which the bees could leave their hives in search of the needful pollen and water.

Winston Churchill has written privately, asking if I would accept a CBE if it was offered to me, for my tracking down of the Piatkow gang last January. I shall decline. I already have my Royal Victorian Chain, and we do not want the house filling up with knick-knacks.

Sunday, 2nd April We filled out the Census form.

Saturday, 6th May From Monday to Wednesday last week I was in London at the Old Bailey, watching the trial of the Houndsditch gang from the public gallery. There were five men in the dock – Yakov Peters and Yourka Dubof, charged with the murder of the policemen, and John Rosen, Karl Hoffman and Max Smoller, charged with the robbery on the jeweller's shop – and two women, Sara Trassjonsky and Nina Vassileva, charged with harbouring felons and conspiracy to rob. The prosecution opened with a confused and confusing speech of two and a half hours. The judge, Justice Grantham, immediately threw out the murder charges, because there was no evidence offered that those two men had ever done any shooting; the real murderers were already dead, shot and burned in Sidney Street. Sara Trassjonsky was found not guilty by the jury, and Nina Vassilieva found guilty only of conspiracy to rob, and given two years. The judge directed the jury to find the three men charged with robbery not guilty, because of lack of evidence. What a farce. I snorted audibly in court several times, and was looked at meaningfully by the Clerk of the Court. *And* Peter Piatkow nowhere to be found. He must have slipped out of the back of the house in Sidney Street, while Svaars and Sokolow were firing from the front.

Wednesday, 10th May May has brought ideal bee-weather, especially in the last few days, and the bees are making the most of it. It has cheered my heart, and the hearts of my fellow bee-keepers in this part of Sussex, enlarged the brood-nests, and increased the population of the hives by leaps and bounds. Though it cannot be said there is abundant forage just at present, the dandelion is beginning to flower in every piece of old lawn and meadow, and the brilliant sunshine has brought the turnips into full bloom. In a few days we shall also have the sycamore, horse-chestnut and beech trees in bloom.

I see the New York gangster Johnny Spanish has been sentenced to seven years. I have followed his career with interest, and that of his partner Nathan "Kid Dropper" Kaplan. Spanish has been the leader for some time of the Five Points Gang, based in the Sixth Ward of Manhattan. I wonder what will happen to the Five Pointers now. I have extensive files on that and the other principal gangs, the Eastman Coin Collectors and the Allen Street Cadets. As well as on Spanish and Kaplan, I also keep notes on Paul Kelly (or Paolo Antonio Vaccarelli, to give him his real name), Johnny Torrio, Edward Eastman, Max "Kid Twist" Zwerbach, Pat "Razor" Riley and several others of the fraternity.

Tuesday, 6th June The continuance of fine, warm weather with the abundant bloom on hawthorn and fruit-trees has put my colonies in fine condition for the main honey flow from the white clover. June promises to be a grand bee month for early honey-gathering, but the continued honey-flow does not conduce to swarming. Every empty cell is used by the workers for storing surplus, thus restricting the brood-nest. The unfavourable month of April depleted the hives by its cold, chilling winds, then came May – all one could wish for – but there was a big leeway to make up to get stocks into swarming condition. It takes a month to breed bees ready to take to foraging, and when the heatwave came, bringing the fields of flowers into bloom a fortnight earlier than usual, the brood combs were used for storing the incoming honey. The brood-combs clogged with the best honey of the year have restricted the growth of the colonies in numbers, so that swarms in this area will be few and late.

Friday, 16th June On Wednesday the seamen went on strike for better wages and conditions, and I believe the dock workers and carters will join them.

Saturday, 17th June I sat out in the garden till very late last night, the temperature being about 70F, and heard what I thought to be four cuckoos. I did not know cuckoos sang at night.

Planted out dahlias, cannas, polyanthus and primroses.

Tuesday, 20th June The Court of Appeal has quashed the conviction of Nina Vassileva, the only person to have been convicted of anything at all at the trial of the Houndsditch gang six weeks ago. Somewhere – perhaps in Riga, perhaps in St. Petersburg - Ulyanov is smiling.

Friday, 23rd June Yesterday King George V was crowned in Westminster Abbey. East Dean celebrated with an unseasonably chilly supper al fresco on the green.

Thursday, 29th June Midsummer day passed, but did not bring us midsummer weather. For a fortnight up to last Thursday we had a dull and sunless period, and as to the bees, alas, I have no good report to make of progress in the apiary. The present honey season opened well with a spell of gloriously fine, warm weather, and with every prospect of a good crowning year to compensate in some measure for the past seasons, in which the bees have done so badly, but the continued heat brought forward the grass crop so rapidly that the farmers cut all our principal bee forage much earlier than usual. The heat also scorched up the young aftermath, and now, since the rain has come, we have only a little sunshine, and low temperatures. The only point in favour of this year – and I grant it is a good one – is that the hives are stocked with food of the finest quality. And the weather in the last few days has improved.

Sunday, 9th July Yesterday I attended a lecture in Eastbourne on the life-cycle of the bee. At one point the lecturer was describing the massacre of drones at the end of the season, and referred to the drones as "males". A near neighbour to me at the back, a woman whose accent showed her to be a Lancashire lass, said, "Sarve 'em reet! Happen we'll ha' votes for women!"

Sunday, 16th July The bees have been in good heart for the glorious days of the first fortnight of July, and the supers have filled as if by magic.

Monday, 24th July The month of July is sustaining its character for "July heat" right to the end, which has caused the bees to cluster in the shady porches of their hives as they do in early June. The limes are in full bloom, and bee work begins early and closes with daylight late in the evening. The continued heat has not started the bees swarming, so that I expect the season for this is now over for this year.

Wednesday, 9th August Today the thermometer recorded 99F. We need rain.

I heard this story yesterday over my garden gate, told to me by Will Stallworthy. Last week a bee-keeper living on Lower Street, on the other side of the village, was preparing to move some hives on a cart. The bees came out, stung the two horses and chased them over several fields, while others attacked the driver and the boy who accompanied him, and followed the bee-keeper inside his cottage where he had fled. The bee-keeper's maidservant ran from the cottage, her pet puppy dog in her arms, and threw herself in the village pond. The bees came out of the back of the house, went on to attack a flock of geese, killing the gander, and then proceeded to attack a cow and her calf, and a tethered goat. The swarm was followed across four fields, and then it came around and back to this village along the main road, and disappeared in the churchyard. It is feared these mutinous bees may be lurking somewhere inside the church, and Sunday congregations may for a while be somewhat smaller than usual. The bees are Carniolans, which are supposed to be as peaceful as my beloved Italians!

Sunday, 13th August Today after the church service Bill Frusher and I went up the steep ladder at the back of the church, through the trapdoor and up into the bell tower, to see if we could see those bees. They were not there, but I did have an opportunity to examine the three old bells. The first has *Sancta Jacobe Ora Pro Nobis* on it; the second, *Byron Eldridge Made Mee 1640*, and the third, *Me Melio Vere Est Campana Sub Ere* – No Bell Under Sky is Better than I.

In the churchyard a little later we saw the tombstone of Parson Jonathan Darby, who died on the 26th October, 1726. It appears that

Darby was a hero of his time – his epitaph is "The Sailor's Friend". At night he used to shine a light from the cliffs in bad weather to warn ships off. The lettering on many of the shapely old gravestones has been erased by overlapping circles of white and yellow lichen.

Wednesday, 16th August In Liverpool the TWF strike has spread to the railwaymen, and there is a general transport strike in that city, except for the movement of essential supplies. The government has lost its senses. It has sent a gunboat to the Mersey, and the army has shot dead two of the strikers. Every port in the country, including London, is now closed. The women in the sweated workshops in Bermondsey, and the cleaners for the London County Council, have joined the strike. This is nothing less than war between Capital and Labour. We have sent our own donation to Tom Mann and Ben Tillett, and collected for the strikers in Newhaven.

Thursday, 24th August This is the fifty-sixth day of the drought, with only a short shower of rain of about fifteen minutes on Saturday, August 5th. We are, therefore, dried up, no water except from the deep wells, and some of the farmers have to fetch water for their cattle from the mill streams several miles distant. And what of the bees? Well, they are retired from business and living on their means which, I am glad to say, are adequate to carry them on for some time. The month of July will be long remembered for its abundant honey-flow during the first three weeks, and the well-filled sections. The quality is fair, but not, of course, equal to the sainfoin and white clover blend of early June.

We are just this afternoon back from three days in Hove, where we watched Sussex play the Indians. It was Fiona's first cricket match, and I am glad it was a gripping contest right to the end. Sussex beat them, but only very narrowly, by 10 runs: Sussex 158 and 149, the Indians 138 and 159. Captain of the Indians was Bhupinder Singh, Maharajah of Patiala, so they have a team of Parsees, Hindus and Muslims, and are led by a Sikh. And their outstanding player is the

Dalit, or "Untouchable", Babaji Palwankar Baloo, who on this tour has taken 114 wickets for an average of 18.8.

Friday, 25th August The Director of the Louvre Museum, M. Guillaume Felix, has written to me, asking me to go to Paris to help with the investigation of the theft of a painting – *La Gioconda*, by Leonardo da Vinci. The painting seems to have disappeared on Monday, though its absence did not cause concern until the following day because staff thought it must have been removed to be photographed. The museum has been closed all this week, and the whole of France is in an uproar. I shall take the ferry from Newhaven on Sunday, and Fiona is coming with me. I imagine whatever useful evidence there may have been at the scene has been obliterated by now, but we may as well have a short summer holiday in Paris.

Sunday, 10th September We are back, but the painting is still missing. My reputation in France will never recover. My command of French proving wholly inadequate to the task of questioning museum staff and the suspects the French police had arrested on suspicion, I was given an interpreter, Mlle. Avril Fox. She is a young English woman, fluent in French, whose father is an attaché at the British Consulate. While Fiona and Mme. Felix enjoyed themselves on the Boulevard Haussmann, Mlle. Fox and I "grilled" – as they say in America – thirty staff in a poky upstairs room at the top of the museum, with very little to show for our time.

On Friday, however, I also questioned two suspects, one a poet and the other a painter, which made the whole, otherwise doomed, enterprise worthwhile. The poet, M. Guillaume Apollinaire, had been arrested on no other grounds than that he once declaimed in a poem, or poetry manifesto, that the Louvre should be burned to the ground. He is just out of his twenties, and an interesting man in that he is both a poet and an art critic. The painter, a Spaniard from Andalusia, is Señor Pablo Ruiz y Picasso, a friend of M. Apollinaire. He was discovered to have in his dwelling an African mask stolen last year from the Louvre, which looked bad for him. The two young men first claimed that they had never seen each other before, and then, when I quickly exposed that ridiculous untruth, each accused the other of stealing the mask. They then suddenly remembered, preposterously, that another man had stolen it and left it with one, or the other, of them, and then they remembered that they did not know it had been stolen at all. It was immediately clear to me that neither of these two young imbeciles had the criminal wit to steal an apple from a fruit stall, and I ordered them to be released at once.

They were so grateful that they took Mlle. Fox and me to a restaurant in Montmartre, where we spent a boisterous evening before going round the corner to the painter's studio. There, under the influence of much red wine, he placed a paper garland on my head, declared me an immortal friend of Art, and gave me a painting on which the paint was only just dry. He said it was a portrait of M. Apollinaire seated at a table having lunch, but as I look at it again now, the only signs of that gentleman's presence are one eye, part of a moustache, and what may be a pipe. There does seem to be a table leg, but no table, and a wine glass, and part of the word "Journal". The rest of the canvas is covered with lines, sharp angles, and fragments that seem to be almost things, but not quite, all in black on a dull ochre and grey palette. It is profoundly exciting, and I like it very much.

I believe I can see a line of progression from the paintings of M. Cézanne to this. Clearly, the conventional mechanisms of perspective and representation are breaking down, but that causes me no alarm. Many conventional mechanisms are feeling under strain at the moment, and not just in the arts. And the traditional function of painting, to fix what one might call an anecdotal fact in comfortingly familiar ways, has also apparently gone out of the window. I think I can cope with that, too. I realise that I am looking at – what shall I say? – one needs to find a vocabulary for this – a new pictorial fact.

I have become distracted from the matter of *La Gioconda*. All I could say to M. Felix when I left was that I believe the thief to be an employee of his, rather than a member of the public or a professional thief, because there was no trace of forced entry, and it would have been very difficult indeed to have got the painting out of the museum during its normal opening hours. I also told him that I believe it is possible that a painting of this kind, which has something of the quality we ascribe to ikons, may have been stolen, not for gain, but for political or religious reasons. In that case it is likely the thief would not harm the work, but would simply wish it to be somewhere else, such as back in Italy, perhaps, or even, given that it could be read as a portrait of the Madonna, in a church. I suggested he watch to see if any of his employees of Italian extraction leave his employment shortly, or take a holiday in Italy.

Señor Picasso is a striking young man. Before we parted he took this photograph from a board on the wall of his studio, signed it with a flourish in blue crayon, and gave it to me. I shall follow the career of this insolent, charming, dark-eyed young painter with interest. As may Mlle. Fox.

Before we left Paris, Fiona and I called on M. Saint-Saëns in the rue de Courcelles. I have been an admirer of his music for some years, and he a generous admirer of my methods of detection, courtesy of the translations of Mme. Bilbault. He gave us tea, specially prepared for us English visitors, and, when I told him of my encounter with the wild young poet and the wild young painter, he confessed to being more and more out of sympathy with the *avant garde* in music. There is no longer, he said, any question of adding to the old rules new principles which are the natural expression of time and experience. What we hear is the casting aside of all rules and restraint. Everyone makes his own rules, he continued. There are no perfect chords, dissonant chords or false chords. All aggregations of notes are legitimate. And this they call the development of taste! I held my peace about my appreciation of the music of Mahler and Schoenberg. On hearing that I keep bees, M. Saint-Saëns showed me an exquisite illustration from the early 1400s, from a work called *Les Très Riches Heures du Duc de Berry*. The illustration is called February, and shows a sublimely beautiful snowy winter scene, including four straw hives, or skeps, in a garden.

Friday, 15th September The month of September brings the bee-keeper's active season more or less to a close, and having secured our crop of honey we must now look forward to what 1912 has in store, and anticipate its coming by getting our stocks into good order. This month I shall examine the brood-combs and, if all is well and they are not clogged with honey, make them snug for the Winter. It is gratifying to learn that the Government has at last allocated a grant for the assistance of bee-keeping. It is quite time that the utility of the bee should be acknowledged as a power in the economy of nature, by reason of the service it renders in the fertilisation of flowers, fruits and vegetables. An experiment made by the late

Professor Cheshire proved conclusively the necessity of insect fertilisation of white clover flowers. A square yard of clover was covered with fine silk gauze, which prevented bees from gaining access to the flowers; this space produced two abortive seeds, while the adjoining square yard, which was not covered, produced more than two thousand healthy seeds.

Sunday, 17th September Yesterday afternoon we had a party of girls and boys from St. Wilfrid's Children's Home, in Shoreham, which used to be a workhouse. They came to see the bees and learn how I keep them, and they stayed for lemonade, tea, cake, jam and honey. They wear a grey uniform with a yellow and black striped tie, and heavy black boots. Why in God's name they have to wear such clothing, I do not know. The boys are taught carpentry and shoe-making, and little else, as far as I can see, and the girls are prepared for domestic service, yet many of the children seem to me just as capable of a liberal education, and just as intelligent, as the privileged children of some of our friends and neighbours. I learned an appalling fact from a boy called Robbie – parents who are in the workhouse can only see their children once every six weeks. I am going to apply myself to getting that cruel rule changed. And many of the children in the home, I was told, are under eight years of age.

Friday, 30th September I have been possessed of a remarkably showy black eye this week. I have not had such a splendid specimen since I was a boy at school. While I was gathering apples last Thursday morning I was stung underneath and a little to the left of my left eye. It did not hurt very much, and I went on with the task. When I came into lunch there was an outcry, and I was led before a looking-glass. I found there was a large claret-coloured patch under the eye. This patch gradually spread all around the eye and all over the eyelid, and remained there for several days, gradually passing away through the usual shades of green and yellow.

Tuesday, 4th October Received yesterday, a gift from M. Apollinaire, in appreciation of my releasing him from the police investigations in Paris a month ago. It is a slightly battered magazine called *Cosmopolis*, which came out in May, 1897. He draws my attention to an extraordinary poem in it by a M. Stéphane Mallarmé, called *Un Coup de Dés Jamais N'Abolira Le Hasard, Quand Même Lancé dans des Circonstances Éternelles du Fond d'un Naufrage* (A Throw of the Dice will Never Abolish Chance, Even When Thrown under Eternal Circumstances from the Bottom of a Shipwreck). The poem spreads itself over twenty pages, in various typefaces and with liberal use of capital letters and blank spaces, with the irregular lines often running back and forth across a double page. These Frenchmen! They are bent on causing the same earthquakes in the arts as Mme. Curie and her colleagues have done in the sciences. Is there something in the water in Paris? Our Laureate, Alfred Austin, must never see this poem. He is in his seventies, and might not survive the experience.

Thursday, 12th October The glorious Summer of 1911, with its long periods of delightful sunshine, has gone at last. As part of my preparation for Winter, I have been examining the amount of food in each of my colonies. I have found at least 30 lbs. of stores in the combs, eight combs in each hive and most of it sealed. This, I am told by Jas. Hiams, will do very well. I have also stretched calico over the hives and painted the material well with oil, which should make them perfectly watertight.

More news from the gang-leader, Arthur Harding. He is in Pentonville. He writes a good letter, and I am always grateful to him for a distant whiff of Hoxton gunfire. His gang has changed its name to the Vendettas. On 10th September, in the Blue Boy in Bishopsgate, the Vendettas attacked the Coons, whose leader is Isaac Bogard. Bogard was cut about the face with a broken beer glass, but survived, and was arrested along with another of his gang. In Old Street Magistrates' Court they asked for police protection, claiming that the Vendettas were waiting for them outside the courthouse. The magistrate, Sir Charles Biron, summoned a vast number of armed police, who arrived at 5p.m. as the two Coons were leaving, their fines paid. A gun battle followed, during which Harding and five others of the Vendettas were arrested. Harding tells me he has retained the notorious Arthur Phale as his solicitor. He writes, "Phale is crooked. That is, you can trust him. Any villainy, he'd do it for you." I suspect Arthur is due a substantial sentence. Our correspondence will, I hope, continue.

Sunday, 16th October This morning my wife and I went with the Mintys to Sponden, in Sandhurst in Kent, at the invitation of General Sir Stanley and Lady Edwardes, who were celebrating their golden wedding. Sir Stanley is a bee-keeper of long standing. In 1858 he joined in the pursuit of Tantia Topee, the most daring and stubborn of Nana Sahib's lieutenants during the Indian Mutiny. He was in the Abyssinian Expedition of 1868, and the Afghan Campaign of 1879-80. I mentioned to Sir Stanley that my good friend John Watson was wounded in that campaign, in the disaster at Maiwand. Sir Stanley said he remembered Watson as one of the most gallant members of the Medical Corps, with an absolute disdain for his own safety where the care of the troops was concerned.

Friday, 20th October Just returned from Cambridge where for three days I have been a guest of Sir J. J. Thompson, of Trinity College. Thompson, of course, won a Nobel prize in 1906 for his discovery of the electron. I wrote to him at that time, congratulating him, and offering some suggestions on possible industrial applications of streams of electrons, or cathode rays, as they are now called. We have continued the correspondence from time to time since then, without meeting, but then Niels Bohr came to Cambridge, and Thompson had the excellent idea that we should all three meet together. We spent our time in almost uninterrupted conversation, on long walks, on the river, at table and in deep armchairs. Bohr, who is still in his mid-twenties, is on top of the world, having gained his Master's degree in April this year, with a ground-breaking dissertation on the electron theory of metals, and he has fallen in love with a charming Danish woman called Margrethe Nørlund. They plan to marry next year. For much of the time, of course, I could not keep up with Thompson and Bohr, but they flattered me by asking for my thoughts on what they were saying, and by referring to a number of my cases in terms that showed they have read Watson's histories. I took this photograph of Bohr and Thompson when we parted, they off to a conference in Heidelberg, and I to Sussex.

Thursday, 26th October We have had continued fine weather, but the barometer is falling steadily, so we are hoping for a plentiful rain to fill the tanks and ponds, as our water supply is running low.

Thursday 2nd November Sussex Bee-Keepers' Association held their first honey show on Tuesday and Wednesday this week at the Dome, Brighton, as part of the Brighton and Sussex Horticultural

Society's Chrysanthemum Show. I entered in the *Six 1lb. Sections* class. There were nineteen entries, and the best I could manage, along with a Miss Vera Hossack of West Wittering, was a Highly Commended.

Friday, 10th November For six days the weather has been appalling. I had to fight my way through hail on the way down to the apiary this morning. All is well there, but we are more or less confined to the house. This has given me time to produce three end-of-year notes, one on honey extraction, one on making candy for the bees, and one on the enemies of the bee. In previous years, with the unavoidable exception of last year, I have appended such notes at the very end of the year, but as I have now written these, I see no reason why I should not include them here.

With the proper resources, extracted honey production is a pleasant pursuit, and the key article of equipment is the extractor. Until the invention of the extractor in 1865 by Major Francesco de Hruschka, of Venice, who one day observed his small son swinging a basket of honeycomb round his head, the nearest approach to extracted honey was strained honey. Surplus honey was removed from the hive by cutting out the combs and mashing them up in a cotton cloth which was hung up in a warm place to drain. Usually, masses of brood and pollen were broken up with the honey, and people simply dug their spoons into the mess of honey and wax, and ate it. With the use of the extractor, frames full of sealed honey are now taken from the hive, and by means of a warm knife the cappings are cut from the comb. (The Bingham knife has of late largely replaced all others.) The frames are then placed in the basket of the cylindrical extractor and the machine started. The centrifugal motion throws the honey out of the comb and onto the side of the can. The frames in the basket are then reversed, and the machine throws out the honey from the other side in the same manner. The honey is then drawn off into galvanised iron cans, or wooden tanks.

Mrs. Holmes's method of candy-making is as follows: she makes about 20 lbs. at a boiling, and uses 10 lbs. of loaf sugar and 10 lbs. of pure Demerara sugar, two quarts of water, and two heaped-up teaspoonsful of cream of tartar. This is put into a saucepan over a bright fire, and kept stirred till all is melted. When it boils, she stands it for twenty minutes over an oil-stove, and this just keeps it boiling, but with no danger of its boiling over. During the boiling, she prepares ten or twelve baking dishes by lining each with a piece of white paper ready for the candy. She then takes a good-sized bath or large pan, and fills this with cold water to come up to the level of the boiling syrup in the saucepan. When the syrup is boiled sufficiently, she stands the saucepan of syrup in the cold water for a few minutes before stirring to cool. At this point, I am allowed to come forward and stir it with a wooden spoon until the syrup becomes creamy in colour and has the consistency of porridge; I then return to the shadows. She pours the syrup out into the dishes, and when they are cold she places over each cake a piece of paper with a hole in the centre to correspond with the feed-hole in the quilt. This quantity of sugar and water makes 24lbs. of good soft candy.

The principle enemies of the bee are, so far as I know, and taking them alphabetically: mice, parasites, and wax moths.

Mice sometimes enter hives in Winter, often by gnawing through an old bottom board, and they may damage or even destroy the colony by eating the combs and disturbing the bees during their winter rest. Hives are best protected from this nuisance by covering wooden bottom boards with thin galvanised iron sheets.

The only parasite I have seen is the Italian bee-louse, on a swarm newly-imported from Italy. I feel safe in saying no fear need be anticipated from them if the bees are kept in strong colonies, and in clean, tight hives, with no old refuse and rubbish accumulating about them.

The yellow larvae, or caterpillars, of the Larger Wax Moth, and of the Lesser Wax Moth, may take over the honeycombs of bee colonies, usually when the bees are in a weakened state. Wax Moths, which are nocturnal and on the wing from June through to August, can often be seen in bee colonies trying to lay their eggs, and in most cases the worker bees will eliminate them and keep the moths from over-running the colony.

It was the good Mr. L. L. Langstroth, just before he died, who showed how spiders may be of use to the bee-keeper. If, he said, spiders have access freely to combs stored in stacked-up hives in the apiary, there never need be any fear that the Wax Moth larvae will be able to do any damage, as the spiders will destroy them.

For the purposes of identification, the adult Larger Wax Moth has a length of just over an inch, a wingspan of the same length and is brown with ash white markings. It can sometimes be seen perching and flying in the vicinity of bee colonies at dusk, usually entering hives or boxes at that time. Females lay clumps of eggs in crevices within the hive, laying between 300 and 600 eggs which are difficult to see. They hatch after five to eight days into the larvae that cause the damage to bee combs. The larvae burrow through combs just under the cappings, leaving a silken tunnel behind them. The bee pupae in the cells are rarely damaged, but sometimes become trapped in the cells by the silk threads and die. The Lesser Wax Moth is less than an inch long, with a similar wingspan, and is silvery-grey to buff in colour, with a yellow head.

Friday, 17th November I have declined the request of the Consett police to investigate the Lintz Green Station murder. My health will not allow me to go to Durham at the moment and, anyway, the trail is cold. On the 9th of this month in the Assize Court the police declined to offer evidence against Atkinson, the man they had arrested for the murder of Joseph Wilson, the station master, and have closed their investigation, for reasons they have not disclosed.

Tuesday, 21st December I have had an early Christmas present from one of Watson's most distant readers. Sent from New South Wales by a Miss Maud Haythornthwaite, it is six saplings of the manuka, or tea tree, which grows uncultivated there, and in New Zealand. Manuka honey apparently has remarkable medicinal properties, particularly in respect of treatment for burns and leg ulcers, and in reducing inflammation. I hope my bees take to its blossoms, when they come.

Sunday, 26th December An excellent Christmas. Watson and Rekha have come down, highly delighted that Rekha has secured a teaching post at the London School of Medicine for Women, under Mrs. Mary Scharlieb, M.D. With Mrs. Trench and Pearl away with their families, we four enjoyed ourselves in the kitchen, with occasional recourse to the claret of which Watson has brought us a case. The result yesterday, at about five in the afternoon, was: purée de pommes Parmentier (which I call leek and potato soup); a rolled mutton joint with capers and anchovies, and a champagne jelly. I long ago used some of my fee in the case Watson called, I think, *The Priory School*, to buy some brandy and port at auction, and I went downstairs after our meal and brought up a bottle of the '65 Danflou. We were very cosy and very happy – they sang *The Twelve Days of Christmas*, to my violin accompaniment – and the wind howled round the house the whole evening. When it was time for bed, and the others had gone up, I opened the back door for a moment to cool my head. I thought for a moment I saw the ghost of Moriarty looking at me from out in the darkness.

1912

Wednesday, 3rd January I look nervously over my shoulder while writing these words, but Mrs. Trench is not about, so I am safe – when scrubbing hives and appliances, a saucepan brush is a useful adjunct to the scrubbing brush. It goes well into corners and along ledges. Common household washing soda, taken from the kitchen, is also an excellent aid. And hints from other trades also go to make bee-work successful. In wiring frames, I have laid the craft of the shoemaker under contribution, and find a fine stab-awl serves admirably for piercing frames, and ¾in. tingles may be used to advantage in fastening wires. The important thing when using a fine awl is to give it a firm, straight twist.

Sunday, 7th January For our anniversary, I gave Fiona a Mah-Jong set of bone and wooden tiles in a wooden case with brass handles. I got it from Sir William Wilkinson, who was Consul-General at Yunnan-Fu and Szemao. He lives in West Dean. For my birthday she gave me *My First Summer in the Sierra*, by the Scot, John Muir.

Indisposition may have prevented me from having a New Year's party, but it cannot stop me being in my laboratory. Wrapped in my herringbone ulster, I have been following my chemical bent, and have been attempting to establish the mineral composition of the bee. I have used only drones found dead last Autumn, dried at 100 to 110 deg., the incinerations being effected at red heat, then the charcoal washed in several changes of water, a small quantity of ashes being the result. The predominant ingredients are sulphur, phosphorus and chlorine.

Sulphur: 1.413 gr.

Phosphorus: 0.953 gr.

Chlorine: 0.294 gr.

Iodine: 0.00009 gr.

Arsenic: 0.00000015 gr.

Silicium: 0.034 gr.

Copper: 0.0006 gr.

Iron: 0.015 gr.

Manganese: 0.002 gr.

Zinc: 0.012 gr.

Aluminium: 0.010 gr.

Calcium: 0.056 gr.

Magnesium: 0.099 gr.

Potassium: 0.025 gr.

Sunday, 14th January Up to the present, conditions this year have been favourable compared with the awful weather at the end of December. Even today I see pollen being carried. Cleansing flights have been frequent. There is fragrant coltsfoot in bloom, and catkins, and Jas. Hiams told me yesterday he had seen a queen wasp in flight in his garden.

Monday, 22nd January Yesterday afternoon I found an apparently lifeless bee on my front doorstep. I took it indoors and held it in the palm of my hand near the fire for a minute or two. At the end of that time, being taken back to the door, it rose and flew straight off in the direction of the hives. I suspect that a large number of bees are lost through these short winter flights on bright days with a cold wind blowing. The bees, after flying about, settle on the ground or on some object near, become chilled and benumbed, and are then unable to return.

Monday, 29th January The Miners' Federation has voted four to one for a national strike, to start in a month's time if their demands are ignored.

Thursday, 8th February I received notice by post today of the sad death in Woodstock, Canada, of Mr. J.B. Hall, one of the most prominent bee-keepers of that Colony, and a correspondent generous on my behalf with both his time and expertise. He brought up a family solely on the proceeds of his apiary, producing in his most successful year no fewer than 25,000 lbs. of honey, and was apparently the soul and wit of the Ontario Bee-Keepers' Association.

Friday, 16th February To Westminster Abbey yesterday for the funeral of the great surgeon, Joseph Lister, as I knew him – 1st. Baron Lister, as the world knows him. He lived not very far from us, in Walmer in Kent, but, to my regret, I never visited him there. He had been sunk in religious melancholy since the death of his wife, and lived as a recluse. He helped me with medical advice twice: once in the affair of the red leech in '94, where he identified the marks, and again in the Cardinal Tosca case, in the following year, when he came with me to Rome to conduct an autopsy of his own.

Saturday, 2nd March Yesterday nearly a million miners stopped work. There is talk among some of my neighbours of martial law, and stockpiling of weapons to be used against the socialist hordes. I know where my own revolver is, and how would use it if I had to. The difficulty in this country is that the workers themselves are more radical than their union leaders, and the union leaders are more radical than the Labour Party, so the workers have no leadership that expresses their anger and organises their strength. Ulyanov has written to me. The British 'proletariat', he says, is changing. He says our government, under pressure from captains of industry who can no longer control their work force, may eventually decide that a foreign war is the only way of dealing with internal unrest. And what, he asks, if the capitalists in Germany, whose interests are essentially the same as those of their British counterparts, decide on the same course? An interesting idea – a war deliberately fomented to sap the revolutionary energy of the labouring classes in both countries. I should dearly like another conversation with Ulyanov.

Wednesday, 6th March Workers are making their way up the alighting-boards with their loads of pollen, gleaned from the expanded crocus blossoms, and betokening the fact that egg-laying has begun. We are now on the threshold of another bee season, one that may, for reasons I cannot at this moment disclose, be my last, at least for a while.

Tuesday, 12th April The bees, and the humans, have been blown off their feet this week. The wreckage of blossom is terrible. Not only has the fruit bloom suffered, but the unopened flowers of the sycamore are strewn about the trees, and the sycamore is an important source of supply in the breeding season. A hailstorm fell this morning that lasted twenty to thirty minutes, and drove us scurrying for shelter.

Monday, 1st April Our little household is in uproar. We are to go to Chicago as soon as travel can be arranged. Fiona's mother, a widow, is in a very serious condition in a hospital there, and may not have long to live. A telegram came last night. Mrs. Trench will look after the house while we are away, Bill Frusher the garden, and Jas. Hiams the bees.

Thursday, 4th April I write in the utmost haste. We leave tomorrow morning on the Newhaven-Dieppe ferry and proceed to Le Havre, where we shall board the *S.S. La Touraine*, the Compagnie Générale Transatlantique's fastest liner, for New York. She sails on the 8th.

Saturday, 1st June We arrived back from Southampton last night. Fiona's mother slowly rallied on seeing her daughter, and when we left Chicago she was back in her own home, with good companions, and a nurse we have contracted to visit her every day.

On the voyage out, after dinner on the evening of Friday, 12th April, I was on the bridge of the *La Touraine*, at the invitation of the captain. He knew who I was from his reading of my adventures in the French translations, and introduced himself at dinner on the first night out of Le Havre. When I told him of my keen interest in wireless telegraphy he kindly gave me the freedom of the bridge to observe the workings of the telegraph and talk with the operators. At about 8.30p.m. that evening we were in contact with the *Titanic*, on her maiden voyage from Southampton to New York. We transmitted to her warnings of field ice and two large icebergs seen in the vicinity of 43° N., 42.8° W, the Grand Banks of Newfoundland. Two days later, as the world knows, *Titanic* struck an iceberg and went down with terrible loss of life. More than 1,500 perished in the icy water.

I inspected the garden this morning, and, of course, the apiary. Every tree is loaded with bloom and the bees are in full work. Jas. Hiams has supered my stocks for me, as the bees have been filling the sections so fast. I do not remember my bees in better condition in June or working more vigorously than at present. Jas. Hiams had a swarm come off on April 20th, he tells me, while he and his wife were out in the garden beating their carpets.

Tuesday, 4th June The month of June has come in in bright sunshine. A few showers of rain in the last two days have been welcome, as Jas. Hiams tells me the whole month of April passed, uncharacteristically, without a single shower. But the temperature is still low for the time of year; even when when we get bright sunshine during the day, the mornings and evenings are cold, and the honeyflow is still in abeyance. Mowing machines – alas! – are laying our sainfoin bloom low.

Wednesday, 5th June M. Apollinaire has sent me another extraordinary magazine, *Der Blaue Reiter Almanach*, published earlier this year in Munich. It is edited by two artists – Wassily Kandinsky, whose show we went to in London three years ago, and a German,

Franz Marc. It has reproductions of more than 140 works, and 14 articles. Yet another direction taken away from our conventional European notions! The works reproduced are largely of primitive, folk, and children's art – there are thirteen pictures by child artists – with pieces from the South Pacific and Africa, Japanese drawings, medieval German woodcuts and sculpture, Chinese paintings, Egyptian puppets, Russian folk art, and Bavarian religious art painted on glass. There are also works by a Dutchman, Vincent Van Gogh, and the Frenchmen M. Paul Cézanne, M. Paul Gauguin, M. Henri Rousseau and – Señor Picasso! He is gaining in confidence, I am sure. His *Woman with Mandolin at the Piano* is quite impossible to read as a normal painting, but has a calm authority. I should very much like to see the original. The magazine also has music – an article by Schoenberg, and the scores of song settings by two Austrian composers, Herr Alban Berg and Herr Anton Webern, both still in their twenties. All this makes me wish for a moment that I were in my own twenties again. I have imagined for a few moments a life dedicated to music.

Sunday, 23rd June Yesterday was a wonderfully close and hot day, and with it came a welcome inflow of honey. The day will undoubtedly encourage many stocks to swarm.

Monday, 24th June I am delighted to enter into my records the case of George Dering, which has just come to light. He was no criminal, but the squire of Lockleys, an estate in Welwyn, where he lived as a recluse until 1879 when he disappeared, more or less, for thirty years. He left instructions with his staff always to keep a mutton chop ready for him, should he reappear, and he did reappear from time to time, always at night, often just before Christmas Day, when he would hastily settle his accounts, pay the staff, read his mail and then leave again within twenty four hours. What has just been discovered, now he has died, is that he has been living quietly all this time in Brighton, under the name of Dale, with a wife and child. The latter have come into a substantial fortune. If only his staff had consulted me. What a pleasure it would have been to investigate this unusual man.

Tuesday, 25th June It has happened. Two of my hives swarmed at the same time yesterday, and joined together on one of Jas. Hiams's apple trees. He got them into a skep, which we weighed, and calculated there were 14½ lbs. of bees inside – in the region of 100,000 of them – the largest swarm he has ever seen.

Saturday, 29th June Yesterday evening old Tavender the blacksmith, now long retired, came to tell me a swarm was out. I followed him to a patch of waste land nearby, where upon the ground and amongst some rough grass a small swarm was clustered. A skep was brought and the operation of hiving successfully performed, but when the skep was lifted for transferring to a more suitable spot, the whole swarm took flight. They clustered once more on the ground a few yards away, but this time on the border of the road. The same process of hiving took place, only to be followed by an encore performance on the part of the bees. They moved slowly along the highway, causing alarm to passing pedestrians, and then into a field, where we followed them. All at once I noticed a change in the direction of their flight. The bees seemed to be heading towards me. In a few seconds I was fairly dressed in bees, from my neck to my ankle down the left side. Gradually these unusually affectionate bees clustered more until the swarm – now about the size of two Association footballs – had collected on my left knee. A curious crowd had gathered on the far side of the hedge to watch. By an effort worthy of a contortionist, I managed to obtain a fair view all round the swarm and discovered the queen on the surface. I picked her off and held her. The skep appeared again, into which I placed my left foot, gave a kick, and then dropped the queen among the bees, which persuaded them to stay there for a breather. Later that evening they were hived by Tavender, but the restless creatures took wing again during the night and went into his allotment shed, where they spent the next two days in a watering can.

Miss Frankenstein of The Hall, Alfriston, told me the other day that a swarm got into a tree outside her bedroom window, about twenty feet high. The longest ladder she had was at least 5ft. too short, but this did not deter Miss Frankenstein. She prevailed on the kindness of the Captain of the Volunteer Brigade, whose fire escape ladder was equal to the task.

Tuesday, 2nd July The last month has been characterised by very unsettled weather, and sections have filled only slowly. However, the past week has seen an improvement. A correspondent in Dumfries, the Rev. W. E. Zehetmayr of the Free Presbyterian Church of Scotland, writes that the bees have been doing badly in the North. Though he has fed continuously, his bees are in a worse condition as regards stores than they were a month ago, though the hives are full – an army waiting for the chance to work. April in the North was apparently a complete failure, and bees had to be fed all the time. May was most erratic, and June arrived accompanied by a hard frost, with an ominous coating of snow on the mountain tops. Mr. Zehetmayr has a colourful turn of phrase. He writes, "our indefatigable bees are desperately trying to reach the moors. Beaten down by the elements, on hands and knees they cross the windswept high road en route to the luring purple heather." Such perseverance (and prose) deserves reward.

We in the South will soon be coming to the end of our honey season, and I am sorry to have to record that only about half a crop has been secured compared with last year, while the quality is not so fine as it was in 1911.

Tuesday, 23rd July For the last two days we have entertained Mme. Curie and her friend and fellow physicist, Miss Hertha Ayrton. Miss Ayrton received the Royal Society's Hughes Medal in 1906 for her experimental investigations into the electric arc. Last night the three of us sat for two hours discussing the non-linear relationship between current and voltage in an electric arc, and the peculiar phe-nomenon of negative resistance as an increase in current results in a lower voltage between the arc's two terminals. When Fiona re-joined us our guests asked for music, so we gave them the Andante, the Allegretto and the Andantino from Schubert's *Fantasy in C.*

Thursday, 15th August This evening in the Tiger I was introduced to a fellow-villager, Mr. Thos. Weatherhogg, who was born on the day after the battle of Waterloo. He has kept bees for over seventy years, using only straw skeps.

Monday, 8th July Yesterday to London – to North End Road, Golders Green, to be exact – to the glamorous house-warming party of Anna Pavlova, the shining star of the *Ballets Russes,* now to live permanently in England. We met many, many Russians, drank in-numerable toasts in vodka, and were introduced in the garden to Pavlova's pet swan, Jack, and her bulldog, Reggie. She gave us this photograph of herself and Reggie, and signed it for us.

Monday, 26th August M. Apollinaire is proving an excellent and most generous correspondent. This morning I had from him a copy of a paper in typescript written by a German mathematician-turned-philosopher, Edmund Husserl. Apollinaire thinks it will interest me. After two hours with the paper on my right and a German-English dictionary on my left I have established that Herr Husserl believes that you can tell when an object occupying your consciousness is a physical thing, by the fact that such an object does not "give itself to you all at once". You can only have one perspective on it at a time, and it invites you to move to one side or the other, or go round the back, to see it from a different perspective each time. The object keeps its unity to itself, and you must construct, or weld it together in your consciousness, creating it for yourself, as one might say, from a hundred different views. My eyes go up to Señor Picasso's portrait of M. Apollinaire, now hanging on my wall, which I imagine would please Herr Husserl, if he could see it.

Thursday, 29th August The weather has been very unsettled for the past fortnight, and the temperature extremely low for harvest time. But there is an abundance of late flowers in the fields and pastures, and a month of fine, warm weather would be a god-send not only to the farmer but to the bee-keeper also, as then our bees would be able to lay in a good store of honey for winter food. Mr. John Moody, who farms over at Friston, tells me the new grasses are a good crop.

Monday, 2nd September Yesterday a jaunt in fine weather to Tunbridge Wells in pursuit of wicker chairs. Excellent oysters for lunch at the Dorset Arms in Withyham.

Sunday, 8th September It gives me inordinate pleasure to report that Mrs. Holmes and I won two prizes yesterday at the Eastbourne Grocery and Allied Trades Exhibition. I took Second Prize in the *Twelve Jars Light Extracted Honey* category, and she took First

Prize, no less, in the *Honey Cake* category. Our kitchen walls are hung with our certificates and rosettes.

Monday, 16th September A letter this morning from M. J. Georges, of 150 rue Bab Souika, Tunis. M. Georges is President of the Tunis Bee-Keepers' Association. He tells me Tunis is an excellent place for bee-keeping, as in many places the pasturage is abundant, and honey produced from rosemary and thyme, being of excellent quality, is in great demand.

Thursday, 7th October I have had a return of the symptoms that troubled me five years ago, and I regret to say that this time they seem to have to have increased in intensity. Dr. Wardleworth has told me that at least three months in a sanitorium are necessary, with controlled diet, specific medicines and complete rest. The place is near Redhill, and there I shall go in a few days' time. To call this prospect unpalatable does not do justice to my feelings, but I shall submit. I have no alternative but to suspend this journal. When I am recovered I shall see if I can arrange for it to be printed locally.

I conclude with some preliminary observations upon the segregation of the queen, made over the last few months. They are incomplete; I hope one day to develop them.

The Segregation of the Queen

Bees construct up to twenty wax queen cells, which are acorn-like and point downwards. The queen lays fertilised eggs in each queen cell. The young (nurse) bees feed the queen larvae with a rich creamy food called Royal Jelly, and extend the cell downwards until it is about one inch in length. Nine days after laying, the first queen cell is sealed with a layer of wax capping. This is the time for a large swarm (called a prime swarm) of bees to leave the hive, led by the older bees. The old queen has been starved of food to make her lighter and able to fly. The older bees cajole the old queen to join the swarm. Eight days later, the first virgin queen leaves her cell. Two things can now occur; either the first virgin queen leads a smaller swarm from the hive, or – as most frequently happens – she locates the other queen cells and kills her sisters by stinging through the wax wall of their cells. About one week later, in the latter case, the young queen takes her first flight to orientate her to her new surroundings. She will shortly take several mating flights in which she can mate with up to twenty drones. Three days later the mated queen will begin to lay fertilised eggs. This queen will stay with the colony until at least the following year, when she too may lead a swarm.

The loss of a queen-bee can occur through accident or deformity, and, unless this loss is remedied in time, the entire colony perishes. In the act of swarming, too, the queen may be lost, and young queens are not seldom lost on their nuptial flight. In the latter case, a young queen may have defective wings which hinder her return, particularly in strong winds, or she may be attacked by birds. Before she leaves she may try to mark well the spot on which her hive stands, by flying with her head towards it for several minutes before rising into the air, yet sometimes, on returning, she misses her mark and attempts to enter the wrong hive, where she is at once seized and killed by the guards at the entrance. This risk is increased when the hives are all alike in shape, are of equal size and of the same colour. Losses from this cause may be minimised by painting the hives in different colours.

Colonies that have swarmed should be examined until the safe mating of the young queen is assured by the presence of eggs and brood, failing which, steps should be at once taken to either introduce a new queen or introduce a comb containing worker-eggs or larvae from which the bees can raise their own queen.

Queen bees reach old age at four or five years, and are generally considered past their prime after their second year. When the fertility of the queen begins to fail, the workers detect it by the lessening of "queen substance" in the hive. Queen substance is a combination of chemicals produced by the queen and licked from her body by her adoring acolytes as she passes among them. As the substance passes from worker to worker, all derive a sense of well-being, and belonging, from it. When the distribution of queen substance falls away, perhaps because of the death of the queen, or her failing powers, the workers immediately supersede her by raising another. In this case the bee-keeper removes the queen, and supplies her place with a young fertile one. It is therefore advisable always to have on hand a supply of fertile queens to meet such contingencies.

When introducing queens, some precautions are necessary, or the bees are likely to destroy the stranger presented to them. If a queen is confined in a cage for a time on one of the combs, and then released, she is generally accepted. Various contrivances have been developed for this purpose, the simplest and most useful being the pipe-cover cage, into which the queen is placed and a card slipped underneath. The cage and card are placed on the comb, where the cage is to be fixed. The card is withdrawn and with a screwing motion the cage is pressed into the comb as far as the base of the uncapped honey cells, taking care not to injure the queen's legs. The bees are then sprinkled with syrup, and a bottle of food placed on top of the hive. In twenty-four hours the hive is opened, the comb lifted out and the cage removed; the best time of day for this operation is early evening, when the bees are resting. At this point, watch the queen carefully; if the bees do not attack her, the comb may be replaced, but if she is seized by the legs or wings, she should be caged again and released in the same way on the following day. Of course, there must not be another queen in the hive during this operation, and any existing queen cells should be removed beforehand.

Queens may also be successfully introduced by means of the travelling and introducing cage. The cage consists of a block of wood with three cavities, in the smallest of which is placed soft candy. A queen and her attendants are placed in the larger cavities, which are in communication with the smaller one, so that when a perforated zinc cover is placed over the openings the bees can get at the candy. A wooden lid is placed over the zinc, and the whole tied and labelled for posting. On receipt of a queen thus packaged, the wooden lid is removed and the cage is put, zinc downwards, on the top of the frames, taking care that the cavities are aligned over the opening between the centre frames. The cage should remain in this position for ten to twelve hours. Then slide the wooden box forward to the end of the zinc, so the food is over the large hole. The bees of the hive will eat through the candy and thus liberate the queen, who will by then have acquired the peculiar scent of that colony. The colony must on no account be disturbed for at least two days after the liberation.

Even these tried and tested procedures are not always successful. When the bees of a colony are intent upon regicide, they usually surround the queen, enclosing her in a living ball so firm and close that it is not easy to break it up. This is known as "balling the queen." A strange queen carelessly introduced, or liberated into a colony being attacked by robber bees, so she is taken for one of the enemy, may quickly be balled or hugged to death. The only sure rescue is to drop the ball into a bowl of water, when it will immediately fall to pieces, the alarm being so great that the murderous design is abandoned, and the queen may be picked out unhurt. If the bee-keeper attempts to break up the ball with the fingers, or with the aid of a smoker, it frequently happens that, when the bees on the outside disperse, one or more of those next to the queen will sting and kill her.

Queens are very easily introduced to swarms. Remove the existing queen from the swarm, shake the swarm from whatever container is being used to house it onto the alighting-board of a hive, and, as they run in, drop the new queen among them. The process is tedious but is quite certain if properly carried out.

I recommend *Queen-Rearing in England*, by F. W. L. Sladen (Houlston and Sons, London, 1905), to those wishing to rear their own queens for the strengthening of colonies, or the creation of new ones.

{PRINTER'S NOTE: The printer wishes to apologise, on behalf of Mr. Holmes, for the somewhat abrupt conclusion to this journal. He takes this opportunity to inform the reader of Mr. Holmes's regret in this regard, and to communicate his own, and Mr. Holmes's, hope that a further volume may be put before the public at a future date, when its author is happily recovered from his indisposition.}

EDITOR'S POSTSCRIPT

What we know, and Sherlock Holmes's readers did not, is that he didn't break off his journal because he was unwell and needed to enter a sanitorium. That story was a blind to cover his real activity in late 1912 when, at the personal request of Prime Minister Asquith, he went undercover as the Irish-American Altamont. In that persona he spent some time infiltrating a Republican Irish secret society in Buffalo, and then in 1913 moved to Skibbereen in the county of Cork, presumably to further strengthen his credentials as an anti-British extremist whom the German secret services in Britain could trust. He would then have come back to this country. By the end of July 1914 he had compromised the entire German spy network in Britain, and thoroughly outwitted its spy-master, von Bork. The *Handbook* must have been printed in Holmes's absence at some time during 1913.

Holmes claimed, in speaking to John Watson in August 1914, that he had started his undercover work in Chicago. I believe that his trip to Chicago in April and May 1912 – recorded in the *Handbook* and ostensibly to visit his wife's mother, who he says was dangerously ill – was also a blind. Holmes and Fiona certainly did go to Chicago at that time, but his main purpose was to lay the groundwork of his new persona as Altamont among the Irish community in that city, either with the Fenian Brotherhood or with Clan-na-Gael.

Why didn't he resume his bee-journal after August 1914? Perhaps he did, but decided not to publish the sequel, or perhaps he published it in a small edition and it hasn't survived. Speculation on the matter is free, but pointless. His papers and possessions – including the Royal Victorian

Chain – were all lost in the fire that gutted Old Home Farm on New Year's Day 1920, as is well known.

Holmes may not have become seriously ill in October 1912 but his general health clearly hadn't been good for some time. It's quite possible that cocaine, tobacco and the relentless activity of the Baker Street years, combined with the stress of more than eighteen months as Altamont, all caught up with him during his sixties, forcing him to give up bee-keeping. What we know for certain is that he became increasingly reclusive during World War I, withdrawing from social life and stopping his international correspondence. He failed to return from a walk alone on the Downs in October 1919, and was never seen again. For a while there was a lurid rumour in east Sussex that Fiona, dressed in black from head to toe and with John Watson at her side, had scattered his ashes on a November dawn in a churchyard on the Pevensey Marshes. Fiona left for America just after Christmas that year. Some see the hand of Emma Moriarty at work in both Holmes's disappearance and the fire. That is possible, I suppose, but two formal enquiries failed to find anything definite at the time. Nearly a century later I think it's safe to say we shall never know how the great man met his end, which of course makes the existence of the *Handbook* all the more precious.

Paul Ashton

Picture Acknowledgements

Grateful acknowledgement is made to the following for permission to use copyright images:

The Margaret Herrick Library of the Academy of Motion Picture Arts and Sciences and by courtesy of the Mary Pickford Foundation (*Mary Pickford*)
Alamy (*Mrs Pankhurst under arrest*)
Arthur Phillips Collection, Museum of Applied Arts and Sciences, Sydney (*man in dark three-piece suit*).
Francis Arcaro (*Antonio Arcaro and ice cream cart*)
Frederick Whitman Glasier, American, 1866-1950
Clowns, circa 1905
Black and white photograph, copy from glass plate negative, 8x10 inches
Negative Number 1130
Collection of the John and Mable Ringling Museum of Art Archives
Lord Egremont (*Lord Leconfield driving a carriage*)
Mirrorpix (*woman on ground*)
Photograph by Horace Warner, reproduced from *Spitalfields Nippers by Horace Warner,* published by Spitalfields Life Books (*two boys, and a single boy*)
Press Association (*Siege of Sidney Street*)
Royal National Lifeboat Institution, Dungeness (*The Lady Launchers)*
© Succession Picasso/DACS 2017 (*Picasso*)
Suffolk Mills Group Collection (*Friston Mill*)
Topfoto:
(*Angela Burdett-Coutts* – Hulton-Deutsch Collections/Corbis Historical)
(*Anna Pavlova*)